LAB MANUAL

Practical Human Geography: Problems, Techniques, Applications

by Paul Hackett, Bram Noble, and Jill Gunn

TO ACCOMPANY

Eighth Edition

HUMAN GEOGRAPHY

William Norton

UNIVERSITY PRESS

8 Sampson Mews, Suite 204, Don Mills, Ontario M3C 0H5
www.oupcanada.com

Oxford University Press is a department of the University of Oxford.
It furthers the University's objective of excellence in research, scholarship,
and education by publishing worldwide in

Oxford   New York
Auckland   Cape Town   Dar es Salaam   Hong Kong   Karachi
Kuala Lumpur   Madrid   Melbourne   Mexico City   Nairobi
New Delhi   Shanghai   Taipei   Toronto

With offices in
Argentina   Austria   Brazil   Chile   Czech Republic   France   Greece
Guatemala   Hungary   Italy   Japan   Poland   Portugal   Singapore
South Korea   Switzerland   Thailand   Turkey   Ukraine   Vietnam

Oxford is a trade mark of Oxford University Press
in the UK and in certain other countries

Published in Canada
by Oxford University Press

Copyright © Oxford University Press Canada 2013

The moral rights of the author have been asserted

Database right Oxford University Press (maker)

First Published 2013

All rights reserved. No part of this publication may be reproduced,
stored in a retrieval system, or transmitted, in any form or by any means,
without the prior permission in writing of Oxford University Press,
or as expressly permitted by law, or under terms agreed with the appropriate
reprographics rights organization. Enquiries concerning reproduction
outside the scope of the above should be sent to the Rights Department,
Oxford University Press, at the address above.

You must not circulate this book in any other binding or cover
and you must impose this same condition on any acquirer.

Library and Archives Canada Cataloguing in Publication
Hackett, F. J. Paul (Frederick John Paul), 1961–, author
Human geography, eighth edition. Lab manual / by Paul Hackett, Bram
Noble, and Jill Gunn.

Includes bibliographical references.
Supplement to: Human geography.
ISBN 978–0–19–544856–6 (pbk.)

1. Human geography—Problems, exercises, etc. 2. Human geography—
Case studies. I. Noble, Bram F., 1975–, author II. Gunn, Jill, author III. Title.

GF41.N67 2013 Suppl.   304.2076   C2013-902959-1

Cover image: Copyright © Elena Elisseeva/Shutterstock

Printed and bound in the United States of America

1 2 3 4 — 16 15 14 13

Contents

About This Lab Manual — iv

Module 1: Understanding Human Geography — 1

Module 2: Space, Place, and Globalization — 13

Module 3: Human Impacts on the Natural Environment — 25

Module 4: Measuring Population Growth — 37

Module 5: Population Patterns and Movement — 51

Module 6: Cultural Landscapes — 63

Module 7: Social Well-Being — 77

Module 8: Political Geographies — 87

Module 9: The Changing Agricultural Landscape — 99

Module 10: Urban Settlement Patterns and Impacts — 115

Module 11: Urban Economies and Transportation — 129

Module 12: Understanding the Industrial Landscape — 143

About This Lab Manual

This lab manual is focused on problem solving and practical applications in human geography. Much of human-geography course delivery is based on the model of "instructor-centred teaching." While that is valuable, we believe that a more hands-on or "student-centred learning" approach is needed so that students can experience human geography by practising it.

The lab manual has three broad learning objectives:

1. To help students develop a variety of essential geographic skills that are grounded in human geography and that they can subsequently build upon in upper-level courses.

2. To enhance classroom discussion and understanding of major geographical concepts and approaches to problem solving.

3. To provide students with "experience" in geographic research and practice in a Canadian context.

The lab manual emphasizes cartographic, qualitative, and quantitative approaches to data collection, problem solving, analysis, and results presentation and interpretation.

The lab manual is intended to accompany and complement William Norton's text, *Human Geography*, but it is entirely suitable for any introductory-level human-geography or general geography course.

The lab manual is Canadian in context. In some cases this means local, place-based field exercises. In other cases exercises focus on differences or connections across space or between regions in Canada, and explore Canada's place in the global economy.

For the student, this lab manual introduces the practical side of human geography—something that is essential to understanding the discipline but that is not usually included in undergraduate human geography textbooks, as these are typically focused on the delivery of fundamental principles and concepts. The lab manual complements textbook- and classroom-based learning by providing students with the opportunity to "do" human geography.

For the instructor, this lab manual reinforces, and adds to, what is being taught in the classroom and will help emphasize the applied side of human geography. Depending on personal preference, the instructor may choose to use the exercises as the basis for a weekly practicum or to substitute a series of lab manual exercises for a major course-based assignment or for a mid-term test, thereby enhancing the practical value of the course. In these regards, the lab manual is intended to be adaptable to the preference of the individual instructor and to the format of the specific course.

The lab manual addresses several themes in human geography, all of which are commonly found in undergraduate introductory courses and each of which is presented here as a module:

- Understanding human geography
- Space, place, and globalization
- Human impacts on the natural environment
- Measuring population growth
- Population patterns and movement
- Cultural landscapes
- Social well-being
- Political geographies
- The changing agricultural landscape
- Urban settlement patterns and impacts
- Urban economies and transportation
- Understanding the industrial landscape

The modules contain exercises designed for students to carry out in the classroom, field setting, and computer lab. Case studies, where used, are designed for both independent reading and thinking and to facilitate in-class discussion. Upon completing all exercises and case-study reviews, students will learn first-hand some basic cartographic skills, how to collect and categorize field data, how to conduct elementary qualitative and quantitative analysis, and how to prepare and assess charts and other graphics designed for presenting and analyzing geographic information. In the end, it is our hope that the student who completes these modules will gain an appreciation of who we are and what we do as geographers, as well, perhaps, as greater appreciation of the nature and diversity of Canadian geography.

Paul Hackett
Bram Noble
Jill Gunn

Department of Geography and Planning
University of Saskatchewan

Module 1
Understanding Human Geography

Introduction

To understand human geography means to understand how societies interpret, place value on, and interact with the landscapes around them. Mapping is at the heart of a human geographer's practice because it helps us identify and communicate about our environment, and in some ways captures cultural and political values about the environment. For example, the W.H. Pugsley Collection of Early Canadian Maps (a holding of the McGill University Libraries) is a set of 50 maps that tell the story of the settlement of North America over a 300-year period (from 1556 to 1857). Each map attempts to depict the location of actual land and water features, but each also conveys a particular political and cultural perspective, reflective of the mapmaker. Historically, geographers focused considerable attention on mapping—charting and describing the earth's surfaces and water bodies (cartography) to aid navigation and wayfinding. But later on, geographers also began identifying geographic patterns (regional and universal geography), and analyzing the complex cause-and-effect relationships of human–land relations (spatial analysis). Today's geographers still do all of these things, and they explore novel areas such as the role that geography plays in globalization, human health, environmental, and cultural phenomena, and much more.

Despite the range and diversity of modern human-geographic thought and practice, human geographers often draw from a common, foundational set of methods, techniques, or tools. Many of the most sophisticated methods in human geography at their core still draw from these foundations. For example, geomatics at its core is still "working with maps" and spatial relations. Many other skills, such as case investigation, primary- and secondary-data collection and analysis, and of course, the ability to think critically about a subject have endured throughout the years and remain as relevant to today's geographers as they ever were. Whether simple or sophisticated, the practice of human geography often involves methods and procedures that help us *make sense of space and place*. To begin this practice you will learn three basic skills in a human-geographer's toolkit: constructing and validating a mental map; field sketch mapping, including working with scale in a field setting; and using the concepts of absolute and relative location to describe landscape features.

Key Concepts

absolute location	ground-truthing	mental maps
map legend	field sketch map	map symbols
relative location	scale	

Learning Objectives and Skill Development

1. To construct a mental map and a field sketch map based on subjective knowledge of a place.
2. To work with mapping legends, symbols, and nomenclature.
3. To understand how to validate or "ground-truth" maps in the field.
4. To understand the concepts of absolute and relative location and distance and be able to apply them in a field setting.
5. To use scaled measurements in a field setting.

Tools Required

compass	graph paper	measuring tape
Internet access	mechanical pencil	coloured pencils

Student Name: _____

Student Number: _____

Course/Section: _____

A. Constructing and Validating a Mental Map

A mental map is the individual psychological representation of space. Mental maps are constructed by having someone draw and label what they recall about their environment on paper or using digital software. Mental maps convey a unique point of view that cannot be captured by any other type of mapping. They can be very useful to a practising human geographer for a number of reasons. For example, mental maps can be used to: find out what features of a person's environment are most significant to them; identify the impact of mass media on perceptions of space and place; understand personal preferences when navigating road systems; and study the efficacy of sign infrastructure to aid wayfinding in public spaces.

1. On the graph paper provided on the next page, draw from memory a map of the grounds that immediately surround the building you are in, taking into account an area that is roughly equivalent to one city block.

   Include the following on your map:
   a. The locations of as many features as possible (e.g., trees and other landscaping elements, roads, pathways, other buildings, structures, and objects).
   b. A map legend to identify features.
   c. A north arrow.
   d. An estimate of the distance in metres between two of the main features on your map (perhaps the distance from the front door of your building to the edge of a road, tree or monument, or perhaps the distance from one corner of your building to another). Note this estimated distance on your map.

2. Once your map is as complete as possible, go out into the field to validate or "ground-truth" some of the features on your map. Compare the features of your map with what is present in the field.

   a. Identify any site features that are missing from your map, as well as any features you included that are not actually present in the field.

   _____
   _____
   _____
   _____

b. Determine the accuracy of your placement of the north arrow. Using a compass, take a north reading. Mark an "X" beside one of the four options below to indicate how well you did:

___ I was correct.
___ I was incorrect and mistakenly labelled east as north.
___ I was incorrect and mistakenly labelled west as north.
___ I was incorrect and mistakenly labelled south as north.

c. Geographers have conditioned us to expect that maps should be drawn with "north as up." How important is it to follow this convention when constructing a mental map?

_____
_____
_____
_____
_____

3. Determine the accuracy of your distance estimate on your mental map. You will learn a simple method of measurement based on the average length of your own stride. This exercise can also be adapted to those whose mobility may be impaired by, for example, estimating the distance covered by one revolution of a wheel.

   To estimate the average length of one of your paces, lay out a 20-metre (m) measuring tape on a section of flat ground, or find an object nearby that is of a known distance. Walk the length of the tape or the object while counting your paces. If it took you 26 paces to walk the distance, each pace would have a length of approximately 0.77 m.

   On grid paper that is divided into 1-cm squares, you can use a 1-cm increment to represent one pace, or 0.77 m. So, for example, if you draw a 5-cm line you are indicating that the line measures approximately 5 paces, or 5×0.77 m = 3.85 m.

   You can "scale up" or "scale down" as needed. For example, you can also decide that a 1-cm increment on your grid paper is equal to two of your paces, or 1.54 m in the field. Thus, if you were to draw a 5-cm line, you are indicating that the line measures approximately 10 paces, or 10×0.77 m = 7.7 m.

   a. Calculate the average length of your pace.

   _____

b. Use your own paces to measure the distance between the two features you selected in question 1c. Record the actual distance below and give two reasons why accurate measurements may be less important when constructing mental maps than for other types of maps.

_____

_____

_____

_____

_____

c. Imagine yourself as a human geographer who has been hired to design a new visitor map for your campus. You survey 100 students on campus and ask them to construct a personal mental map of the campus, including buildings and pathways. How could including both "too much" and "too little" information on a mental map compromise its usefulness to you?

_____

_____

_____

_____

_____

B. Field Sketch Mapping Using Scale

Field sketch mapping is an important mapping-based skill but differs from mental mapping. It involves constructing a representative map based on close observation within a bounded area. The purpose of field sketch mapping is to record the location and label features of interest in the field as accurately as possible. Field sketch mapping is useful for practising human geographers who need to develop an up-close, in-depth personal understanding of an area whose features are not otherwise readable through any other type of map (aerial, satellite, etc.). Field sketch maps are "context specific," meaning that they depict only the features deemed important by the mapmaker. Urban planners can use field sketch maps of intersections, for example, to understand why people avoid crossing on a certain side of the road and recommend needed improvements to infrastructure. Field sketch maps:

- Are drawn from "a bird's-eye view."
- Provide an up-to-the-minute representation of what is actually on site.
- Allow researchers to make their own decisions about what details they wish to record on a map and those they wish to omit.

- Include a map legend and notations about what is observed.
- Are based on physical interaction with a site, which is often critical to understanding and solving a research problem.

Field sketch maps need to be clearly drawn, labelled, and annotated. This does not mean they need to be works of art, but they should be neat, clearly labelled, and undertaken with a view to enhancing your ability to identify and analyze key site features in the field once you have returned to the classroom. It is important to be as neat and as accurate as possible because very often field sketch maps are used as communication tools or to recall vital field information.

1. Using the grid paper on the next page (with 1-cm increments), choose a small outdoor site as the basis for your field sketch map (e.g., a building site, your backyard, a small park). Place a title at the top of your map that describes the location.

2. Decide on the boundaries (perimeters) of your site and draw them. To make this process easier, divide the boundary into segments and measure them one at a time. Use the length of your average pace (see question A.3.a.) as a basis for calculating distance. Now draw the site boundary to scale on your map. You will adopt a 1:100 map scale, which means that every unit on the map is equivalent to 100 units on the field. Thus, a 10-mm (1-cm) increment on your grid paper represents 1000 mm (or 1 m) in the field. If you take a field measurement of 7.5 m, for example, you would represent this on your map with a 7.5-cm line. Indicate the map scale you are using (1:100) at the bottom of your map.

3. Indicate the location of the site's major interior features on your map, relying on your paces as a form of measurement and the map grid as a guide for correct placement.

4. Develop a unique symbol to represent each new class of features on your map, which might include benches, light poles, trees, etc. Organize these symbols into a map legend (list of symbols with explanations) that appears beside or on a blank section of the field sketch map.

5. Add a north arrow on your map, using your compass.

6. Recall question A.3.c. With the project's purpose in mind, list two advantages and two limitations of the geographer making his or her own field sketch map rather than collecting mental maps from the public.

_____

_____

_____

_____

C. Absolute and Relative Location

Absolute location is the absolute position of a place or site in space; it is usually described by giving an exact physical location, such as a street address or geographic coordinates. Absolute location is often enough information if a person is already familiar with the context of the place or site, or has access to a GPS system. Describing its location in relation to other places or land features can also convey the location of a place or site. This is a referential method of describing location and is often used when absolute location is not known, and to assist in navigational or wayfinding situations.

1. List three to five human-made or naturally occurring features that could be used to describe a relative location. For example, one human geographer making land use maps of remote transmission line rights-of-way in northern British Columbia used the steel tower numbering system to indicate the relative location of certain types of land uses such as unmarked snowmobile trails. A simpler example would be referring to the location of a youth camp at a lake to describe the relative location of a nearby cabin inaccessible by road.

   _____
   _____
   _____
   _____
   _____
   _____
   _____

2. Using Google Maps or a similar online application, find the coordinates of the Canadian landmarks and monuments listed in the table below. Search for the exact latitude and longitude for each site (degree, minute, second). Record these "absolute locations" in the following table:

Site	Geographic Coordinates (degree, minute, second)
a. CN Tower, Toronto, ON	
b. BC Place Stadium, BC	
c. Iqaluit airport, NU	
d. Signal Hill, NL	
e. Your university campus	

3. List two ways to describe absolute location other than by geographic coordinates. For example, a street address is also an absolute location.

   _____
   _____

4. Describe the absolute and relative locations of the following, <u>without</u> using geographic coordinates:

Site	Absolute Location	Relative Location
a. the nearest town or city hall		
b. your current place of residence		
c. your current university or school		

5. Identify three cases or situations in which it would be necessary, or preferable, to describe places or sites using absolute location.

_____

_____

_____

6. Identify three cases or situations in which relative location would be more commonly used to describe places or sites.

_____

_____

_____

Module 2
Space, Place, and Globalization

Introduction

Geography is often referred to as the "spatial discipline." For geographers, the term spatial refers not only to the way things are distributed across the earth's surface, but also to the way movement occurs and the way human processes operate. Geographers have had a long-standing interest in why things are located where they are, how activities and actions in one region influence those in another, the nature of interaction across space, and the flow of people, goods, ideas, and technologies.

That said, some of the fundamental principles of geography, namely location, distance, and movement, have been significantly redefined in recent years as the various processes of human activity become increasingly global in scope. This new condition, often referred to as globalization, is also defined as a process—a set of worldwide processes that make the world, its economic system, and its society more uniform and more integrated. An important question that geographers often face, then, is whether geography still matters in a globalizing world. If we are indeed living in what has been referred to as a global village, is distance still important? Does location still matter?

In the first part of this module you will apply a number of geographic techniques to examine patterns of movement and the influence of distance on movement and spatial interaction. Then, you will be challenged to think critically about whether these fundamental principles of location and distance are still relevant in a globalizing economic system.

Key Concepts

spatial interaction	gravity model	critical distance
frictionless zones	time-space convergence	globalization

Learning Objectives and Skill Development

1. To apply geographic models to analyze movement and interaction.
2. To examine the concepts of location, distance, and interaction.
3. To examine critically the role of geography in a globalizing economy.

Tools Required

calculator	a map of your local city or region

Student Name: _____

Student Number: _____

Course/Section: _____

A. Spatial Interaction

Newton's law of universal gravitation states that the attraction or pull between two objects is proportional to the product of their masses and inversely proportional to the square of the distance between them. In other words, big objects attract each other more than do smaller objects, objects that are in proximity have a stronger mutual attraction than objects that are farther apart, and attraction between any two objects decreases as the distance between them increases. In human geography, this attraction or gravitational pull is simply the quantity of movement between two objects, or places.

In mathematical terms, the **gravity model** can be expressed as follows:

$$I_{ij} = K \frac{P_i \times P_{ji}}{D_{ij}^b}$$

where

I_{ij} is the interaction or movement between two locations, i and j

K is a constant that adjusts gravity model estimates such that expected and actual interaction is approximately equal

P_i and P_j are the sizes, usually population, of locations i (origin) and j (destination)

D_{ij} is the distance between locations i and j

b is a distance decay function, typically from 0.5 to 3, where the higher the value the greater the friction of distance (i.e., distance becomes more of a constraining factor on movement)

Later in this manual you will examine population and movement. In this exercise you will examine critically the gravity model in action as a tool to predict and understand migration flows.

The table on the following page presents population, distance, and actual migration data for each Canadian province and territory with respect to the destination province of British Columbia. For this exercise you will be using a simplified version of the gravity model:

$$I_{ij} = K \frac{P_i}{D_{ij}}$$

Note that the destination (P_j) has been omitted from the table because you will be examining a single destination, British Columbia, and the distance decay function (b) has been omitted for simplicity. The K-value has been set at 2.68 to calibrate the model.

Population, Distance, and Migration for Each Canadian Province and Territory with Respect to the Destination Province of British Columbia

Origin province (i)	Origin population (P_i)*	Distance (D_{ij})	P_i/D_{ij}	Predicted migration (K-adjusted)	Actual migration**	Residual (%)
AB	3,815,498	460			22,099	
NT	43,730	790			401	
SK	1,068,117	795			2,969	
YT	35,604	870			494	
MB	1,259,375	1,170			3,645	
NU	33,608	1,370			56	
ON	13,438,807	2,110			13,306	
PQ	8,011,996	2,320			2,571	
NB	755,835	2,670			764	
NS	948,459	2,800			1,001	
PE	145,855	2,820			411	
NL	513,503	3,200			493	

*Based on Statistics Canada, Table 051-0005, "Estimates of population, Canada, provinces and territories, quarterly (persons)", CANSIM (database), using E-STAT (distributor). (Q1 2012)

**From: Statistics Canada, Table 051-0019, "Interprovincial migrants, by province or territory of origin and destination, annual (persons)", CANSIM (database). (2011/12)

1. Calculate column 4 in the table above by dividing the population of the origin province by the distance between the origin province and British Columbia.

2. Calculate the predicted migration between the origin province and British Columbia by multiplying your results for P_i/D_{ij} by the K-value 2.68.

3. The residual is the difference between the actual migration and the predicted migration, or how well the gravity model predicted migration flow. To calculate the residual, subtract the predicted migration from the actual migration and divide the result by the actual migration. Multiply by 100 to convert the residual to a percentage. This value can be interpreted as the percentage error of the gravity model. For example, −4.8 means that the actual migration was 4.8% less than the predicted migration, as a percentage of the actual migration.

4. What is the effect of the population size of the origin province and the distance between the origin province and British Columbia on *predicted* migration?

5. As shown by the actual migration, what is the significance of distance?

6. What were the most *over-predicted* and *under-predicted* migration flows?

7. Are there factors other than physical distance that should be considered so as to improve the predictions of the gravity model? (Hint: What factors might explain the errors in prediction for the two most over- and under-predicted migration flows to British Columbia?)

B. Critical Distance

Most people make far more short-distance trips than long-distance ones. If you were to draw a circle on a world map that encompassed the farthest location you have travelled from your current location, you would note that the number of trips you have taken to the boundary are far fewer than the more daily, routine trips within a much smaller boundary around your current residence.

This zone of daily or routine activity is referred to as the *frictionless zone*. The boundary of the frictionless zone is referred to as the *critical distance*. Beyond this critical distance, an individual's movement or frequency of travel decreases because of cost, distance, effort, and means.

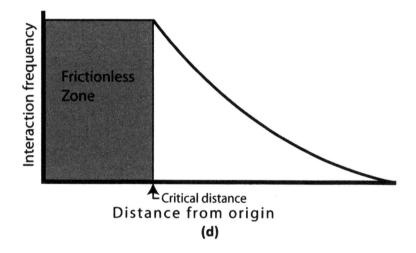

(d)

1. On a map of your city, town, or region, draw a circle (or a semicircle, depending on the structure of your built environment) that roughly depicts your "frictionless zone"—the area in which you move about regularly and in which your movement is not constrained or affected by time, distance, cost, and other factors. Use your current residence or campus as the centre point.

2. Using the scale on the map, determine the total *area* of your frictionless zone. (*Hint: The area of a circle* = πr^2.)

3. What are the main factors or constraints that define your personal frictionless zone and critical distance?

   _____
   _____
   _____
   _____
   _____

4. Compare your frictionless zone and critical distance to those of your classmates. Identify the different factors that influence an individual's frictionless zone and critical distance.

   _____
   _____
   _____
   _____
   _____

C. Time-space Convergence

Places are separated on the landscape by *absolute distance*. However, as communications and transportation technologies improve, the *time distance* between two places diminishes. People and commodities can now move from one place to the next much faster than before. Today, we often tend to measure distance in terms of time, rather than kilometres. For example, if you were flying from Calgary to Toronto, would you describe the distance as "approximately 2,700 kilometres" or as "approximately 4 hours"?

This reduction in travel and communication time between two places is, in many cases, similar to a reduction in the absolute distance between places. This principle is referred to as *time-space convergence* or compression.

The time-space convergence between two places can be calculated as a rate:

$$TSC = \frac{\Delta TT}{\Delta T}$$

   where

   TSC is "time-space convergence"
   ΔTT is the difference in the travel time between places in time period n and time period $n - 1$
   ΔT is the difference between time period n and $n - 1$

Example:

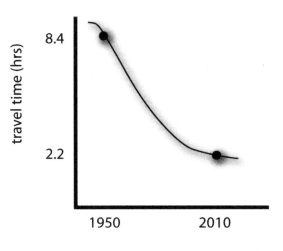

Travel time from location A to location B in 1950 was 8.4 hrs.

Travel time from location A to location B in 2010 was 2.2 hrs.

$$TSC = \frac{2.2 - 8.4 \; hrs}{2010 - 1950 \; yrs}$$

$$TSC = -0.10 \, hrs/yr$$

1. Calculate the time-space convergence between Halifax, Nova Scotia and London, England.

Date	Travel Time
1920	340 hrs
1960	8.5 hrs
2013	5.8 hrs

Time period	TSC
a. 1920 to 2013	
b. 1920 to 1960	
c. 1960 to 2013	

2. Why do you suppose the TSC rates are so different between the time period 1920 to 1960 and 1960 to 2013?

   _____
   _____
   _____
   _____
   _____
   _____

3. In 1992, it took approximately 6.5 hours to travel by car from Toronto to Ottawa. With improvement to transportation infrastructure, the time in 2013 is only 5 hours. What is the time-space convergence?

4. Calculate the TSC rates for past and present travel from London to Edinburgh.

Date	Travel Time
1770	6,000 min.
1860	700 min.
1950	200 min.
2013	90 min.

Data source: Based on Janelle (1968) and British Midland Airways (2010).

Time period	TSC
a. 1770 to 1860	
b. 1860 to 1950	
c. 1950 to 2013	

5. What do you think were the main advances in transportation during each of these time periods that affected TSC rates?

D. Diffusion

When people move or migrate they often carry new ideas, clothing styles, or even new technologies with them. The physical spread of such ideas, cultural traits, innovations, and even diseases through the movement of people, such as migration, is referred to as **relocation diffusion**. However, phenomena can also spread in many other ways including hierarchically, contagiously, and through stimulus.

Hierarchical diffusion involves the trickling down of phenomena from the socially elite or from large nodes or centres to the rest of society. For example, a new idea or style may first appear in New York and quickly become adopted in Toronto, then Vancouver and eventually trickle down to Edmonton and to other smaller centres. **Contagious diffusion** refers to the rapid spread of phenomena due to direct exposure or personal contact. The spread of disease is perhaps the most commonly cited example of contagious diffusion, but the contagious diffusion of ideas, news or trends also occurs through social media. **Stimulus diffusion** is when a new idea, innovation or cultural trait is adopted by another person or society but is given a new or unique form.

1. Identify the type of diffusion that best describes each of the following examples:

Type of diffusion	Example
	A new migrant family from Germany brings with them a well-known home cooking recipe that quickly becomes popular amongst the community association in your local neighbourhood.
	Your friend shows up to class on Monday wearing the latest style of jacket that you saw worn by a celebrity on the Hollywood Academy Award television show on the weekend.
	The Toronto Maple Leafs National Hockey League team makes it to the Stanley Cup finals. Within a week almost everyone you pass on the street, even those who are not hockey fans, is now wearing a Toronto Maple Leafs jersey.
	Apple was the first computer company to adopt the idea of clicking on an icon to open a software program. IBM, whose operating system initially required the user to type in a command prompt before opening a program, quickly adopted it.
	A new McDonald's restaurant is opened in India. To serve the local Hindu population, the restaurant manager replaces the meat on the menu in order to serve veggie burgers.

	You post a video of your cat playing the piano on YouTube for your mom to see and soon you have millions of viewers from around the globe sharing their comments.

2. Provide an example, different than those in the previous question, for each of the following types of diffusion:

Type of diffusion	Example
Relocation diffusion	
Hierarchical diffusion	
Contagious diffusion	
Stimulus diffusion	

E. Does Geography Still Matter?

Do the above principles of distance and spatial interaction still apply? Does time-space compression mean that distance no longer matters? Is location and place still relevant in a globalizing economy characterized by high-speed travel, e-commerce, and nearly instant communication? Over the past 30-plus years, many people have said that distance, location, and place are no longer relevant—geography is dead!

In his 1970 book *Future Shock*, for example, Alvin Toffler writes about "the demise of geography," suggesting that "thanks to transport and communication technologies, and flows of people and information, geographical difference is being dissolved." In 1992, Richard O'Brien also claimed the "end of geography," arguing that location no longer matters (O'Brien, 1992), and in 1997 Frances Cairncross wrote about "the death of distance" (Cairncross, 1997). An article appearing in *Fortune Magazine* in 1998 similarly proclaimed "the end of geography"—Gary Hamel and Jeff Sampler wrote that with the rise of e-commerce, "customers, as well as producers, will escape the shackles of geography."

1. In this exercise you are challenged to take a side and think critically. Adopt a perspective and write a critical essay to defend your position: "geography is alive" or "geography is dead." Were these people right? Does geography still matter? Provide examples to support your claim of how geography does or does not matter. You should draw on examples from communications, global trade, culture, or the location of industries or businesses to support your position.

References

British Midland Airways 2010. http://www.flybmi.com.

Cairncross, F. 1997. *The death of distance: How the communications revolution will change our lives.* Cambridge, MA: Harvard Business Press.

Hamel, G. and J. Sampler. 1998. The E-corporation. *Fortune,* 138 (11), 80–93.

Janelle, D.G. 1968. Central place development in a time-space framework. *Professional Geographer, 20,* 5–10.

O'Brien, R. 1992. *Global financial integration: The end of geography.* New York: Council on Foreign Relations Press.

Toffler, A. 1970. *Future shock.* New York, NY: Bantam.

Module 3
Human Impacts on the Natural Environment

Introduction

Relationships between humans and their physical environment are often at the centre of human geography studies. Up until the second industrial revolution in the mid-19th century, the impact and geographic extent of human settlement were limited and relatively harmless: world population was much smaller then, and most modern-day chemicals and transportation and production technologies did not exist. The human population has grown exponentially over the past century and a half, and as societies have become more technologically advanced (particularly in the northern and western hemispheres) there has been a corresponding rise in the pressure on the resource stores that fuel our economy and the natural systems that sustain them. The vastness of the North American land base and its relatively sparse interior and northern population, particularly in Canada, has led to the common misconception that supplies of natural resources are "endless," accompanied by aggressive resource-development policies. Vast hinterland regions were, and often still are, viewed as storehouses of nature's bounty rather than integral ecological systems. The cumulative drawdown of resource supplies and often unfettered dumping of society's wastes are now linked with many serious regional environmental issues, from deforestation, wildlife extinction, and habitat destruction to water and air pollution, ecosystem simplification or collapse, and possibly global climate change. Environmental stewardship programs and sustainable-development policies have become a top scientific and governmental priority since the late 1980s, and we are now learning to work together to slow, repair—and prevent—the harmful environmental impact of urban and regional development on sensitive environmental receptors, such as water, soil, air, vegetation, and wildlife.

This module focuses on common property resources. In particular, the exercises examine the implications of common property resource management regimes through two case studies. In the first exercise, you will examine changes to Newfoundland and Labrador marine fisheries. You will work with a data set, construct graphs that depict trends in resource harvests and value, and analyse those trends. In the second exercise, you will learn about cumulative impacts to common property resources, taking eutrophication of the Great Lakes as an example. You will observe changes to pond water samples to which nutrients have been added to better understand the process of cultural eutrophication, and later consider the value of scientific experiments to human geographers. In the final exercise, you will reflect upon these two cases in the context of managing "common property resources." You will learn about a phenomenon known as the "tragedy of the commons" and what it means for environmental management and decision-making.

Key Concepts

cultural eutrophication	Great Lakes	east coast fisheries
tragedy of the commons	sustainable development	cumulative effects

Learning Objectives and Skill Development

1. To develop an understanding of common property resources.
2. To learn how unmitigated use or stress can lead to cumulative effects and resource collapse.
3. To understand how science can be applied to environmental management and how environmental knowledge can be increased through case investigation.

Tools Required

Internet access
two large samples of pond water

8 clear plastic containers
access to low wattage fluorescent lighting

Note to Instructor

This module is best delivered across two class meetings. At the first meeting, perform Part A of the module and the first part of the experiment outlined in Part B. At the next meeting, complete the experiment outlined in Part B, as well as Part C.

Student Name: _____

Student Number: _____

Course/Section: _____

A. Case Study 1: East Coast Marine Fisheries

The following table contains data on marine fish landings (i.e. total catch or harvest) and the total economic value of the landings, from 1986 to 2000, for the province of Newfoundland and Labrador. Data are also provided for two different types of fish: groundfish, which includes such species as Atlantic cod, Halibut and Pollock; and shellfish, which includes such species as Crab, Shrimp and Lobster.

Year	Total Landings (metric tonnes)	Total Value (thousand dollars)	Groundfish Landings (metric tonnes)	Shellfish Landings (metric tonnes)	Groundfish Value (thousand dollars)	Shellfish Value (thousand dollars)
1986	517,646	209,993	399,075	18,179	144,746	33,307
1987	502,360	291,859	395,096	21,904	215,854	43,273
1988	560,786	291,449	383,332	39,702	167,292	82,829
1989	521,829	265,934	347,813	43,289	155,534	75,632
1990	545,500	285,506	336,560	47,259	175,260	78,252
1991	427,133	263,482	270,868	48,948	148,913	88,888
1992	285,291	198,351	156,199	57,320	85,049	93,924
1993	247,600	209,095	97,650	67,060	50,882	118,429
1994	140,067	232,201	35,370	78,524	19,177	196,973
1995	141,763	349,299	22,189	91,406	19,866	313,559
1996	195,347	289,759	27,542	108,562	27,184	233,539
1997	220,287	325,530	39,335	131,313	34,093	260,688
1998	265,052	386,255	51,666	145,893	60,368	299,932
1999	276,456	533,044	68,078	162,950	84,173	424,800
2000	275,268	586,887	67,822	161,866	79,596	488,238

1. Construct a graph depicting the trend in "Total Landings" in the Newfoundland fishery from 1986 to 2000. You can construct your graph manually on the following page or input the above data table into Excel to generate the graph. If you generate your graph using Excel, make sure to label your axes.

2. Construct a graph depicting the trend in "Total Value" in the Newfoundland fishery from 1986 to 2000. You can construct your graph manually on the following page or input the above data table into Excel to generate the graph. If you generate your graph using Excel, make sure to label your axes.

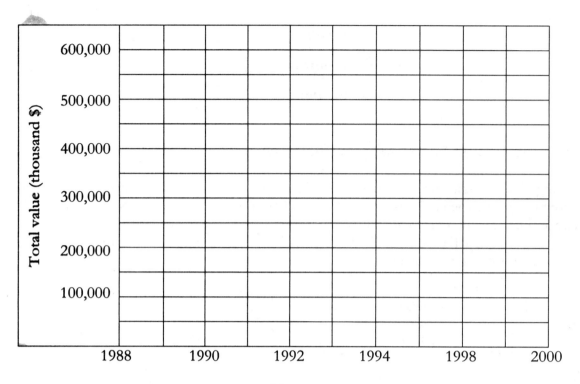

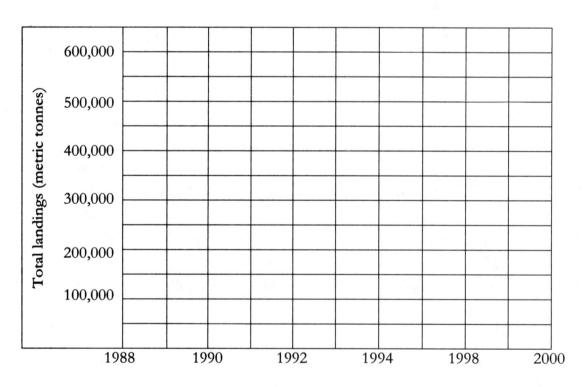

3. Use your graphs to describe the trends in Total Landings and Total Value in the Newfoundland and Labrador fishery from 1988 to 2000.

4. What happened in the Newfoundland and Labrador fishery in the early to mid-1990s to cause such a sharp decline in Total Landings?

5. Despite Total Landings remaining relatively low, the Total Value of the Newfoundland and Labrador fishery increased significantly in the late 1990s. Why do you think this was the case? (Hint: Examine the trends in Groundfish and Shellfish landings and values in the table on page 27).

_____

_____

_____

_____

B. Case Study 2: The North American Great Lakes

The pollution of the Great Lakes was noted as a serious problem as early as the 1950s. By the 1970s, Lake Erie was known as the "Dead Sea of North America" due to a process called "eutrophication." This is the process by which lakes, ponds, and streams become enriched with inorganic plant nutrients, particularly phosphorus and nitrogen. Eutrophication is a natural process, but it is sometimes accelerated or intensified by the land fertilization activities of humans. Fertilizer runoff from agricultural lands, such as those around Lake Erie, can lead to increased levels of phosphorous and nitrogen and therefore increased biological productivity. Without proper policies and regulations to control the total amount of inorganic nutrients from all sources reaching water bodies, eutrophication can result in the proliferation of devastating algal blooms that make a lake unappealing to swimmers and boaters. "Algal mats" can form and cover a lake completely. If this happens, the lake is slowly "suffocated" due to reduced oxygen levels in the water. Over time, eutrophication can cause serious stress to commercial and sport fisheries, or their complete loss.

The following experiment simulates human-induced (also called "cultural") eutrophication. It is similar to a series of experiments conducted in a remote region of Canada whereby large-scale ecological manipulations were performed to learn more about how added carbon, nitrogen, and phosphorus affected rates of eutrophication. These experiments resulted in changes to water quality protection laws and reduced nutrient loading, allowing some lakes to fully recover. In this part of the module, you will first make a prediction and validate it based on observation. You will then consider how physical and predictive science can help inform the work of human geographers.

Your instructor will begin the experiment by obtaining water samples from two different ponds—one that is quite eutrophic, and one that is not. These samples should be placed in large, clear containers so the water is easily observed.

1. Observe the two water samples and make notes on the following for each (supplement observations with sketches as appropriate):

    (a) water colour

(b) turbidity (cloudiness or haziness caused by suspended solids)

(c) presence of biological organisms

(d) water odour

2. (a) Write a short statement predicting which water sample contains more nutrients based solely on your observations (nitrates and phosphates for example).

_____

_____

_____

(b) State which observations the prediction is based on.

_____

_____

_____

Next your instructor will simulate cultural eutrophication by adding fertilizer to the water samples. **Instructor**: Divide each of your two large pond water samples into four equal parts using clear containers (e.g. plastic soda bottles with labels removed). Set aside a sample of each kind of pond water to act as a "control." To the remaining three pairs of samples, add 5 ml, 10 ml, and 15 ml of liquid plant fertilizer, respectively. Place all eight samples under low wattage fluorescent lighting for a day or more, until the class meets again. Keep the samples away from sources of bright light such as windows. Then ask students to observe the changes.

3. Observe all eight water samples, making notes on the following (supplement observations with sketches as appropriate):

   (a) water colour

   (b) turbidity (cloudiness or haziness caused by suspended solids)

   (c) presence of biological organisms

   (d) water odour

4. (a) Was your prediction correct about which kind of pond water was more eutrophic?

   _____
   _____
   _____

   (b) Reflecting on what you observed, explain why environmental management initiatives (such as a joint government program to clean up the Great Lakes) are often "reactive" rather than "proactive" in nature.

   _____
   _____
   _____
   _____
   _____

5. Cultural eutrophication is a phenomenon caused by the incremental or cumulative effects of many human actions over time to a common property resource. Because cumulative effects are often individually insignificant, they tend to be missed or ignored in environmental planning and management. List five factors that you think can hamper communication among scientists, the public, and the government regarding common property resources, and/or avoidance of cumulative environmental effects.

___

6. Reflecting on the Great Lakes case and your experiment results, why might it be important for human geographers to understand, consult, and/or use predictive science in their work?

___

C. Managing Common Property Resources

In his now classic paper, originally published in 1968, Garrett Hardin (1968) described the "tragedy of the commons" and the conditions that arise when there are insufficient limits to growth. Search your library first to see if you can locate a copy of this article. The original version was published in *Science* and is available online at http://www.sciencemag.org.

If you are not able to obtain a copy from *Science*, you may be able to find a copy of the article using the Google or Google Scholar search engines.

After reading Hardin's article, describe:

1. What is a "common property resource"?

2. What are the major challenges to managing common property resources or protecting them from human impact or overuse?

3. Do the cases addressed in this module (the east coast marine fisheries or the North American Great Lakes) depict any of the characteristics of what Hardin describes as the "tragedy of the commons"? Explain your answer.

D. Questions for Classroom Discussion

1. What does the concept "tragedy of the commons" have in common with the concept of "sustainable development"?

2. What kinds of government policies or regulations might be needed to manage common property resources in a more sustainable manner? Focus on freshwater or fisheries as examples.

References

Contant, C. & Wiggins, L. 1993. Toward defining and assessing cumulative impacts: Practical and theoretical considerations. In S. Hildebrand & J.B. Cannon (Eds.), *Environmental analysis: The NEPA experience* (pp. 336–56). Boca Raton, FL: Lewis.

Cornell University. "Eutrophication experiments." Environmental Inquiry. Ithaca, New York. Accessed online [6 May 2013] at: http://ei.cornell.edu/watersheds/Eutrophication_Experiments.pdf

Groffman, P., Baron, J., T. Blett, Gold, A. Goodman, I., Gunderson, L., Levinson, B., Palmer, M., Paerl, H., Peterson, G., Poff, N., Rejeski, D., Reynolds, J., Turner, M., Weathers, K. & Wiens, J. 2006. Ecological thresholds: The key to successful environmental management or an important concept with no practical application? *Ecosystems, 9*(1), 1–13.

Hardin, G. 2009. The tragedy of the commons. *Journal of Natural Resources Policy Research, 1*(3), 243–253. Originally published in *Science, 162*(3859) (December 13, 1968), 1243–1248.

Mack, J. "Eutrophication." Lake Scientist. Accessed online [6 May 2013] at: http://www.lakescientist.com/learn-about-lakes/water-quality/eutrophication.html.

Fitzpatrick, J., Di Toro, D. "A History of eutrophication modelling in Lake Erie." HydroQual, Inc.: Mahwak, N.J. Accessed online [6 May 2013] at: http://www.ijc.org/php/publications/html/modsum/fitzpatrick.html.

Module 4
Measuring Population Growth

Introduction

At its core, demography is a specialized discipline which studies human populations and their basic characteristics. Among other topics, demographers ask questions about the composition (age, ethnicity, and gender) and size of the population, how quickly or slowly it is growing, and how its numbers are affected by births and deaths (Beaujot and Kerr, 2004, p. 3). They may also focus on inherently geographical questions, such as differences in the density of populations between regions (spatial variation), and the magnitude and impact of, and rationale for, migration.

Population geography is a branch of geography that is related to demography in that it also aims to understand and measure certain characteristics of human populations. However, it does so with a particular focus on the variations in these characteristics across space or, as in the case of migration, on how these processes play out across the earth's surface and over time. Population geographers are also interested in how local conditions at a given place affect local demographic variables, and vice versa.

The immediate subject of interest, and consequently the approach, of the population geographer can vary considerably, from the basic to the complex. Thus, geographer Wilbur Zelinsky identified three approaches, each with increasing complexity (Zelinsky, 1966). At the most basic level, a researcher would simply describe the population at a given location in terms of numbers and other demographic characteristics. More advanced studies would identify and explain the observed spatial patterns of these phenomena. Finally, Zelinsky noted that, at its most advanced, population geography seeks to understand the differences in population that can be observed across space, and the contribution to those differences made by local characteristics or phenomena (pp. 5–6).

In this module you will explore population geography through a series of basic exercises, designed to introduce you to a few of the concepts and techniques employed in this branch of geography. To complete the two parts of this module you will create, assess, and compare population pyramids; and evaluate the age and sex composition of a population.

Key Concepts

population pyramid	age cohorts	sex ratio
dependency ratio	youth dependency ratio	old-age dependency ratio
population structures		

Learning Objectives and Skill Development

1. To understand population structure and composition.
2. To construct and interpret population pyramids.
3. To calculate sex and dependency ratios for First Nations and provincial populations.
4. To use population statistics to compare population groups.

Tools Required

calculator

Student Name: _____

Student Number: _____

Course/Section: _____

A. Population Pyramids

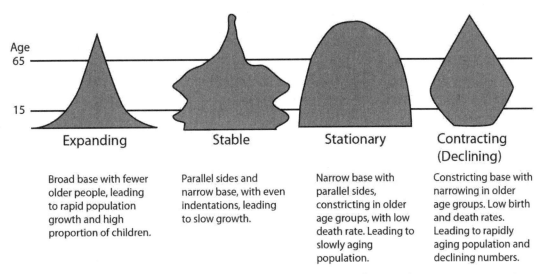

Source: Adapted from http://www.statcan.gc.ca/kits-trousses/animat/edu06a_0000-eng.htm

In this section you will construct three population pyramids based on historical and current data from a Manitoba First Nations community. You will then analyze and compare these pyramids to determine whether or not the composition of this population has changed over the past century. Finally, you will construct a pyramid from current provincial data (for Manitoba), and compare this to the current First Nations pyramid.

A population pyramid is a type of chart, shaped somewhat like a pyramid, employed by demographers and others to depict graphically the structure of a population. There will likely be more information on population pyramids in your text, along with examples, but if there is not, you can visit the Statistics Canada site for further information (Statistics Canada, 2010).

Essentially, a population pyramid is composed of two horizontal bar charts, one each for females and males, placed back to back, with percentage of the total population (or number of individuals) measured out along the horizontal axis, and age along the vertical axis. This provides a total view of the age and sex structure at a given moment in time, but it also records the demographic impact of historical events, such as mortality from past wars, or past trends, such as culturally induced increases or declines in the birth rate. By evaluating the shape of these pyramids, we can determine the general trends of the population, including whether it is in a period of growth or decline, or whether it is aging or growing younger, as well as the ratio of males to females for any given age group (cohort).

1. Island Lake population pyramid

The people of Island Lake belong to four distinct First Nations (Garden Hill, Red Sucker Lake, St Theresa Point, and Wasagamack), which together form the Island Lake Tribal Council. The four communities are located in northeastern Manitoba and are all fly-in communities, with limited road access via temporary winter roads for a few weeks each year. Although they are independent First Nations and differ somewhat from each other in family interrelationships, culture, and economy, they have been treated by the Canadian government as a single unit for much of the period since signing a treaty in 1909. Consequently, early population counts were collected for a single Island Lake community, and while this has changed in recent decades, for the purposes of this exercise we will treat them as a single population.

Begin by constructing pyramids for Island Lake for the years 1901, 1969, and 2005, using the data provided in the table on the following page. Although it is possible to construct these either manually (by hand), on the graph paper provided for instance, or digitally, using a program such as Excel, your instructor may have a preference. So please consult him or her before beginning.

For this part of the exercise you will want to follow a few basic guidelines.

- Use five-year age groups, or cohorts. Note that the data have already been aggregated into these cohorts.
- The upper age cohort will be 70+.
- Males will be on the left and females on the right.
- Place the vertical scale (y-axis) on the left side of the pyramid, rather than in the middle.
- For the horizontal scale (x-axis) measure the increments as percentages of the total population.

Island Lake Population

Age Cohort	1901		1969		2005	
	Male	Female	Male	Female	Male	Female
0–4	39	47	304	276	571	569
5 to 9	42	43	224	240	495	420
10 to 14	40	33	190	190	444	438
15 to 19	30	20	141	130	412	397
20 to 24	25	28	122	128	339	303
25 to 29	23	25	102	97	291	283
30 to 34	17	15	68	83	252	250
35 to 39	11	15	41	47	243	215
40 to 44	6	14	42	43	193	180
45 to 49	6	8	41	33	163	146
50 to 54	3	10	29	25	124	110
55 to 59	1	2	26	23	86	86
60 to 64	2	5	17	15	76	69
65 to 69	2	4	22	17	42	41
70+	4	5	30	23	47	58

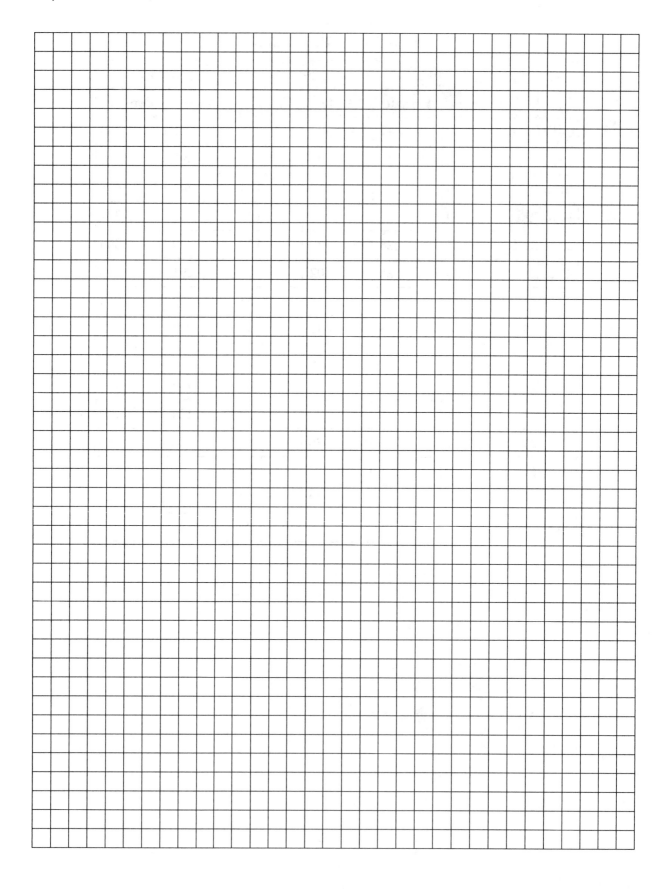

After completing these pyramids, answer the following questions:

a. Describe the Island Lake population pyramids for 1901, 1969, and 2005; i.e., what shape is each, and what does this tell you about the Island Lake population at that time? Please note that the names used for types of pyramid can vary from source to source. If your text does not identify these types, and your instructor does not have a preference, you may want to use the terms favoured by Statistics Canada, i.e., **expansive**, **stable**, **stationary**, and **declining** (Statistics Canada, 2010).

Year	Description of population pyramids
1901	
1969	
2005	

b. How did the population structure of Island Lake change from 1901 to 2005, if at all?

_____

_____

_____

_____

_____

_____

2. Manitoba population pyramid

Although the Island Lake people live within the province of Manitoba, their population is subject to different trends and forces than is the general population. In the second part of this section you will create a population pyramid for the province of Manitoba using 2011 data (Government of Canada, 2011) (see table below). Use the same conventions for designing and evaluating your pyramid as indicated previously. The graph paper is located on the next page.

Manitoba Population, 2011

Age Cohort	Male	Female
0–4	39,275	37,905
5 to 9	38,380	36,240
10 to 14	40,960	38,395
15 to 19	44,175	42,035
20 to 24	41,935	40,990
25 to 29	38,900	39,285
30 to 34	37,000	38,265
35 to 39	37,475	38,100
40 to 44	38,455	39,320
45 to 49	44,695	45,395
50 to 54	44,945	45,030
55 to 59	39,350	40,415
60 to 64	33,725	35,170
65 to 69	24,405	25,840
70+	50,875	71,340

Data Source: https://www12.statcan.gc.ca/census-recensement/2011/as-sa/fogs-spg/Facts-pr-eng.cfm?Lang=eng&GK=PR&GC=46

a. How does the most recent Island Lake pyramid (2005) compare to that of Manitoba (2011)? If they differ, explain how they are different and why this might be the case.

___
___
___
___
___
___

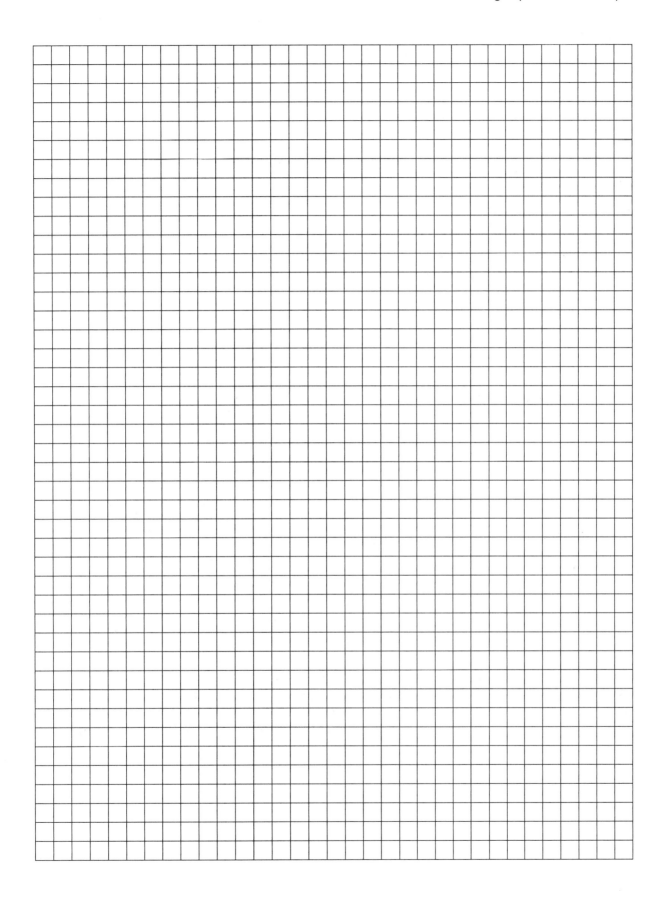

b. What does the shape of Manitoba's pyramid imply for future population growth in the province?

_____

_____

_____

_____

B. Dependency Ratios

The dependency ratio (DR) is a measure of the portion of a population that falls outside of the productive cohort, which is generally designated as those people aged 15 to 64. Those who are 14 or younger, or over 65, are considered to be economically dependent on the remainder of the population. There are actually three dependency ratios: the **youth dependency ratio** (YDR); the **old-age dependency ratio** (ODR), and the **total dependency ratio** (TDR). For this exercise you will be calculating all three of these DRs, which are derived from the following formulas:

$$TDR = \frac{(\text{\# people aged } 0-14) + (\text{\# people aged} \geq 65)}{\text{\# people aged } 15-64}$$

$$YDR = \frac{\text{\# people aged } 0-14}{\text{\# people aged } 15-64} \times 100$$

$$ODR = \frac{\text{\# people aged} \geq 65}{\text{\# people aged } 15-64} \times 100$$

In each case the resulting DR is then expressed as a percentage of the productive cohort. In general, the higher the TDR of a given population, the higher the burden on the working-age group. Where the TDR is greater than 100, there are more dependent people than productive. Should the TDR be too high, that population may face some economic difficulty in providing for all members. Using the data (for 2012) in the table on the next page, calculate the YDR, ODR, and TDR for Canada and for the provinces and territories and answer the questions that follow.

Provincial and Territorial Populations by Age Cohort, 2012

	Age Cohort			YDR	ODR	TDR
	0–14	15–64	65+			
Canada	5,634,906	23,865,707	4,983,362			
Newfoundland & Labrador	76,014	355,437	81,449			
Prince Edward Island	23,278	99,511	22,906			
Nova Scotia	138,873	652,998	156,587			
New Brunswick	113,711	518,451	123,173			
Quebec	1,242,711	5,482,023	1,253,255			
Ontario	2,202,338	9,266,795	1,897,161			
Manitoba	234,916	841,745	175,029			
Saskatchewan	201,065	701,265	155,474			
Alberta	690,051	2,677,180	410,841			
British Columbia	685,836	3,190,006	700,735			
Yukon Territory	6,055	26,210	3,133			
Northwest Territories	9,422	32,238	2,552			
Nunavut	10,636	21,848	1,067			

Source: 2012 Data http://www.statcan.gc.ca/tables-tableaux/sum-som/l01/cst01/demo10a-eng.htm

1. Which of these jurisdictions would appear to have the largest potential problem with an excessive TDR? Which of the two dependent cohorts, the young or the elderly, contributes most to their TDRs?

2. What are some possible economic implications of having a high YDR? A high ODR?

_____

_____

_____

_____

_____

3. Why might two countries having the same TDR but much different values for YDR and ODR have very different long-term prospects?

_____

_____

_____

_____

_____

4. In 2000 Canada's ODR was 16.9. How does this compare to the ODR for 2012 that you calculated above? What does the difference between these two years say about the long-term trend for Canada's population?

_____

_____

_____

_____

_____

C. Sex (Gender) Ratios

The sex, or gender, ratio (SR) is a measure of the balance between males and females in a given population. It is calculated with the following formula and is expressed as x number of males for every 100 females:

$$SR = \frac{\#\ males}{\#\ females} \times 100$$

Overall, we would expect the SR for any given population to be approximately 100, with roughly equal numbers of males and females. However, the SR value will naturally vary from an even division for the youngest and oldest cohorts. For instance, it is normal for slightly more males than females to be born (about 105 males per 100 females) (Trovato, 2009). This changes gradually over the lifespan, such that following the age of 60 females increasingly outnumber males. As well, other forces may intervene to influence the SR artificially for a given cohort.

Referring to the population data for Newfoundland and Labrador data (table below), calculate the SR values for each cohort in 2012 and enter them into the table.

Newfoundland and Labrador Population, 2012

Age Cohort	Male	Female	SR
0–4	12,910	12,033	
5 to 9	12,463	11,814	
10 to 14	13,665	12,836	
15 to 19	14,983	13,821	
20 to 24	16,171	15,581	
25 to 29	14,313	14,961	
30 to 34	14,559	15,176	
35 to 39	14,866	16,242	
40 to 44	17,431	18,511	
45 to 49	20,427	21,257	
50 to 54	21,003	21,502	
55 to 59	20,484	21,657	
60 to 64	19,201	19,792	
65 to 69	14,980	15,727	
70+	24,259	30,034	

1. Describe the general pattern of SR in Newfoundland and Labrador's population.

_____
_____
_____
_____
_____

2. What factors might account for discrepancies in the SR in the middle cohorts of countries in the developing or developed world (not only in Newfoundland and Labrador) in favour of either males or females?

___

References

Beaujot, Roderic P. & Kerr, D. 2004. *Population Change in Canada*. Don Mills, ON: Oxford University Press.

Government of Manitoba, Health Information Management Branch. 2009. *Manitoba Health and Healthy Living Population Report*. Winnipeg. http://www.gov.mb.ca/health/population/pr2009.pdf. Accessed 16 May 2010.

Statistics Canada. 2011. http://www12.statcan.gc.ca/census-recensement/2011/as-sa/fogs-spg/Facts-pr-eng.cfm?Lang=eng&GK=PR&GC=46. Accessed 13 May 2013.

Statistics Canada. 2012. http://www5.statcan.gc.ca/cansim/a26?lang=eng&retrLang=eng&id=0510001&paSer=&pattern=&stByVal=1&p1=1&p2=37&tabMode=dataTable&csid=. Accessed 13 May 2013.

Statistics Canada. 2010. http://www.statcan.gc.ca/kits-trousses/animat/edu06a_0000-eng.htm. Accessed 16 May 2010.

Trovato, F. 2009. *Canada's population in a global context*. Don Mills, ON: Oxford University Press.

Zelinsky, W. 1966. *A prologue to population geography*. Englewood Cliffs, NJ: Prentice-Hall.

Module 5
Population Patterns and Movement

Introduction

In the previous module we looked at a few ways of analyzing and describing the demographic structure of a community or region. While population pyramids provide a snapshot of that structure, and dependency and sex ratios are useful tools for assessing age and sex balance, we may also want to look for broader patterns that allow for a more ready comparison between populations—patterns that reflect the nature of settlement in those areas. At the same time, such patterns are not static, and they may be influenced by a number of different demographic processes. Foremost of these are births, deaths, and migration, the movement of people from one jurisdiction or administrative area to another.

In this module you will investigate one particular type of demographic measure, population density, which can be a useful tool for comparing settlement patterns and population in different countries or other geographic entities, particularly those of unequal size, for which direct comparison of absolute numbers might be misleading. As well, you will look at the migration process, a key determinant of population change over time. To complete the three parts of this module you will calculate and map population density in Saskatchewan by census division; calculate migration rates for select Canadian provinces and evaluate historical trends in migration between them; and carry out an interview with a real or imaginary migrant in order to explore the importance of push and pull factors in migrants' decisions.

Key Concepts

population density   census divisions   interprovincial migration
population change   choropleth map   immigration rate
emigration rate   net migration rate   internal migration
push factors   pull factors

Learning Objectives and Skill Development

1. To understand and apply the concept of population density as a tool for comparing the demographic attributes of different places.
2. To define migration and to understand the differences between internal and international migration.
3. To document the reasons why people migrate from one place to another.

Tools Required

calculator   pencil or coloured pencils for shading

Student Name: _____

Student Number: _____

Course/Section: _____

A. Population Patterns: Mapping Density in Saskatchewan

It is understood that population patterns vary across space. We know intuitively and by experience that, for instance, some countries or provinces have larger populations than others. Thus, India has an estimated population in excess of a billion, while Monaco has only perhaps thirty thousand people (US Census Bureau, 2009). However, the absolute population will tell only part of the story. Algeria (estimated 34,178,188) and Canada (33,487,208) have similar populations, but Algeria is only about one-quarter the size of Canada. In this case, we can readily determine that the overall **density** of Algeria, i.e., the number of people per unit of land, is far greater than that of Canada. Determining that allows us to draw some conclusions about the nature of life in Algeria as compared to Canada.

By the same token, however, we know that population density can vary significantly within countries. So, for example whereas Canada's population density according to the 2006 census was 3.5 persons per km^2, Ontario's was 13.4 and Saskatchewan's only 1.6 (Statistics Canada, 2009a). Moreover, there are also substantial variations in density within the provinces and territories, the most densely populated areas being found within a relatively short distance of the border with the United States. Nevertheless, as a device for comparing populations between jurisdictions, density is easily calculated, and it allows for a more meaningful comparison of population when the spatial extent of the communities is markedly different.

In this section you will calculate and map population densities for Saskatchewan by census division. Once completed, this map will give you some insight into the nature of settlement in different parts of Saskatchewan at the last census. Begin by completing the table on the next page, filling in the columns for **% change** (2006 to 2011) and **2011 population density**. For density, the formula is simply the population divided by the area.

Saskatchewan Population by Census District, 2006, 2011

Census Division	Population (2006)	Population (2011)	% Change	Area (km²)	2011 Pop. Density (persons per km²)
Division No. 1	29,168	31,333		14,997	
Division No. 2	20,363	22,266		16,859	
Division No. 3	13,133	12,691		18,554	
Division No. 4	11,086	10,879		21,365	
Division No. 5	30,529	32,007		14,783	
Division No. 6	220,688	237,746		17,548	
Division No. 7	45,532	46,648		18,836	
Division No. 8	29,199	29,962		22,233	
Division No. 9	34,736	35,314		15,271	
Division No. 10	17,680	17,546		12,220	
Division No. 11	244,273	271,170		16,683	
Division No. 12	22,452	23,228		13,887	
Division No. 13	22,342	23,089		17,255	
Division No. 14	36,515	37,195		33,818	
Division No. 15	79,018	83,725		19,613	
Division No. 16	37,118	37,845		21,828	
Division No. 17	40,406	44,180		22,458	
Division No. 18	33,919	36,557		270,068	
Total Population	968,157	1,033,381		588,276	

Next, take the values for population density for each census division and map them on the attached base map. Here you will be creating a choropleth map, a common form of thematic map in which areas are shaded, with the intensity of the shading (brightness) determined by the magnitude of the attribute being mapped. In this case the attribute is population density, the darker shades representing higher densities.

As with the pyramids, it is possible to create this map either manually (by hand), shading in the attached map with coloured pencils for instance, or digitally, if you have the skills and the software. Since your instructor may have a preference, please consult him or her before beginning. In either case, attach the completed map to your assignment. In case it accidentally becomes detached from the rest of this assignment, remember to put your name, student number, class number and the date neatly on the map. (If you wish to work with the digital version, it can be found at http://www.statcan.gc.ca/ca-ra2006/m/sask_cd-dr-eng.pdf.)

Instructions

Step 1	Prepare the data by dividing the values into classes. Rather than selecting a distinct shade for each census division, you will be arranging the divisions into groups, or classes, based on the density. For this exercise you can use quartiles to classify the data. Detailed instructions for creating quartiles can be found at the Statistics Canada site (Statistics Canada, 2009b).
Step 2	Select an appropriate shading scheme. You will need four shades of the same colour, or of grey, to represent the four classes. Remember that the shades must all be of the same hue, or colour, that they must be clearly distinguishable on the final map, and that they must follow a logical progression, with lighter shades representing lower density and darker shades representing higher density.
Step 3	Apply the correct shades to the census divisions.
Step 4	Finally, add any other necessary information to your map, including the legend, data source (Statistics Canada), and the personal data noted above.

After completing the map, answer the following questions:

1. In general, what broad geographic patterns emerge with respect to population change in Saskatchewan between 2006 and 2012? For this question refer to your percentage change column and the map.

   _____
   _____
   _____
   _____

2. What general pattern of population density in Saskatchewan can be seen on the completed map?

   _____
   _____
   _____
   _____

3. What can be observed about population density in census divisions with urban centres?

   _____
   _____
   _____
   _____

Population Patterns and Movement | 55

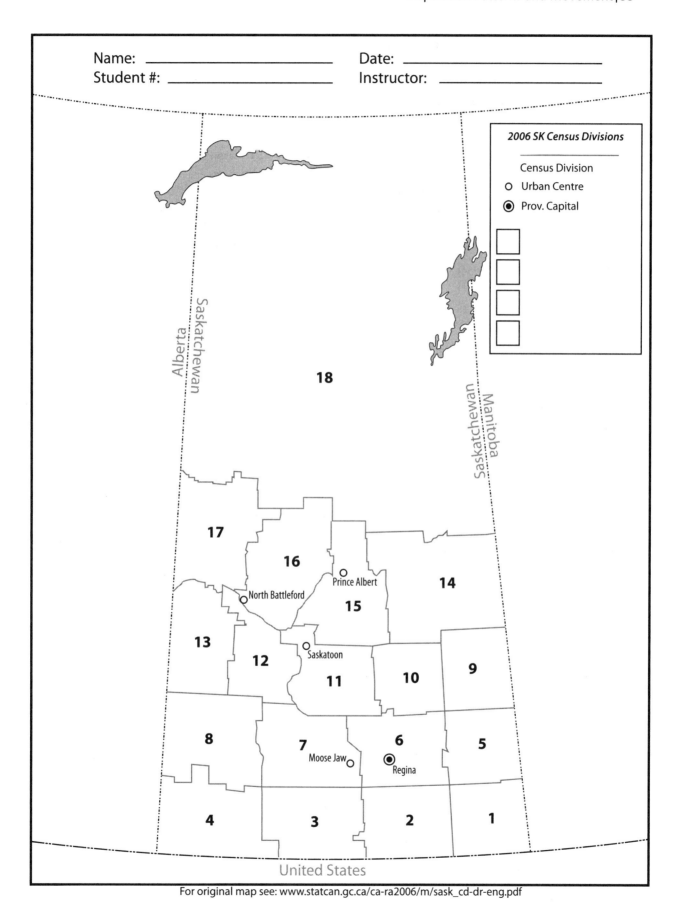

B. Interprovincial Migration in Canada

We could define migration as relocation from one place to another, but although simple, this definition is too inclusive since it embraces all sorts of human movement that we would not consider migration. We can distinguish migration from other forms of mobility, such as a change in neighbourhood, the daily journey to work, or a visit to another country, by virtue of the fact that, to be considered migration, such movement must pass three tests (Trovato, 2009, pp. 332–3). First, there must be a permanent or semi-permanent change of residence. Secondly, it must be a movement across long distances. Finally, migration generally crosses major administrative boundaries. In a sense, this movement occurs in two directions: people come to new places and they also leave old ones. In international migration, those leaving are called emigrants (and they emigrate to a new country), while those who are arriving are called immigrants.

However, migrations can also occur within countries; this is referred to as **internal migration**. In this case, the terminology is modified slightly. For internal migration the immigrant becomes the in-migrant, and the emigrant becomes the out-migrant. As noted above, migration is a crucial force in population change, and in some circumstances the presence or absence of migration will determine whether a population grows or not. This may be as true of the province as it is of the country.

Alberta-Saskatchewan Migration, 2000–2009

Your task is to examine the trend in interprovincial migration between the neighbouring provinces of Saskatchewan and Alberta during the first decade of the 21st century. Traditionally, this flow has favoured Alberta; for over the last two decades of the 20th century net migration was 76,690 in the westward direction (232,340 from east to west, and 155,650 from west to east) (Encyclopedia of Saskatchewan, 2006). This may have been due to the decline in the farm population in Saskatchewan as well as the province's fluctuating economy, combined with the strength of the Alberta economy. However, in recent years this has shifted, with an increase in the eastward flow across the Alberta-Saskatchewan border.

Begin by graphing the two flows of inter-provincial migration based on the table on the next page. Create a line graph with both series represented on the attached graph paper (page 58). Be sure to label the axes and provide an accurate title. Once the graph is completed, answer the questions that follow.

Interprovincial migration between Alberta and Saskatchewan, 2000 to 2009

Year/Quarter	Direction of Migration	
	SK to AB	AB to SK
2000/03	3,026	1,477
2000/06	3,936	2,147
2000/09	3,584	1,908
2000/12	2,682	1,177
2001/03	2,832	1,145
2001/06	3,382	1,627
2001/09	4,257	2,101
2001/12	2,650	1,380
2002/03	3,244	1,424
2002/06	4,027	2,151
2002/09	4,012	2,142
2002/12	1,960	1,475
2003/03	2,687	1,420
2003/06	3,139	2,430
2003/09	3,048	2,066
2003/12	1,932	1,194
2004/03	2,392	1,809
2004/06	3,371	2,209
2004/09	3,656	1,932
2004/12	2,695	1,106
2005/03	3,595	1,428
2005/06	4,530	1,860
2005/09	3,955	1,977
2005/12	2,993	1,074
2006/03	3,600	1,556
2006/06	2,683	1,872
2006/09	3,716	2,702
2006/12	2,145	2,082
2007/03	2,052	2,470
2007/06	2,230	3,326
2007/09	3,111	6,022
2007/12	1,986	3,036
2008/03	2,738	3,449
2008/06	3,391	3,930
2008/09	2,833	3,849
2008/12	1,871	1,917
2009/03	3,041	2,567

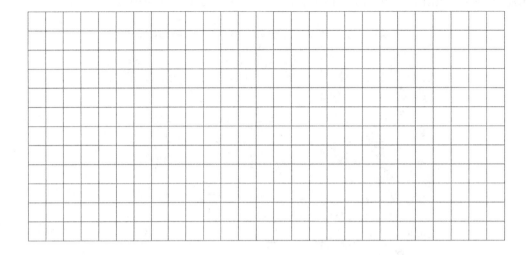

1. At what point during this period did migration from Alberta to Saskatchewan first exceed migration to Alberta from Saskatchewan?

2. What factors may have played a role in the recent shift in population movement in favour of Saskatchewan?

3. What impact would the movement to Alberta of large numbers of people in their productive years have had on the economy of Saskatchewan?

Alberta Migration Rates 2009

One way to gauge the numerical importance of migration to a place is by calculating the migration rate. This is the number of migrants arriving at a destination over a given time period, per 1,000 population. In fact, there are three migration rates, and each measures a slightly different aspect of migration. The immigration rate (IR) measures the rate of migrants moving to the place. The emigration rate (ER) measures the rate of migrants leaving the place. Finally, the net migration rate (NMR) yields the difference between these two. The formulas for these are as follows:

$$IR = \frac{Number\ of\ immigrants}{Total\ destination\ population} \times 1,000$$

$$ER = \frac{Number\ of\ emigrants}{Total\ origin\ population} \times 1,000$$

$$NMR = \frac{Number\ of\ immigrants - Number\ of\ emigrants}{Total\ population} \times 1,000$$

In this exercise you will calculate the NMR for Alberta (for the fourth quarter (Q4) of 2012, or October 1 to December 31) by using the above formula and the data provided on the next page in the Interprovincial Migration table. Note that you are given data for both interprovincial and international migration, and so will have to aggregate these figures to calculate the NMR.

After calculating the NMR, answer the following question.

1. What is the NMR for Alberta for Q4 2012? How does this compare to the NMR for Canada from Q2 2009 (2.51 per 1,000 people)?

_____
_____
_____
_____
_____

Interprovincial migration to and from Alberta, Q4 2012				
Destination of Emigrants (from AB)		NMR	Origin of Immigrants (to AB)	
NL and Labrador	437		NL and Labrador	606
Prince Edward Island	8		Prince Edward Island	175
Nova Scotia	257		Nova Scotia	1,414
New Brunswick	222		New Brunswick	805
Quebec	252		Quebec	1,298
Ontario	1,838		Ontario	6,382
Manitoba	620		Manitoba	986
Saskatchewan	1,532		Saskatchewan	2,166
British Columbia	3,118		British Columbia	5,762
Yukon	15		Yukon	135
Northwest Territories	66		Northwest Territories	181
Nunavut	12		Nunavut	0

Alberta's estimated population for 2012: 3,815,498
International Immigrants to Alberta, Q4 2012: 9,081
International Emigrants from Alberta, Q4 2012: 1,567

Source: Data compiled from Statistics Canada (2009) and Alberta (2012).

C. Understanding Migration: A Case Study

Generally speaking, the decision to migrate long distances is not taken lightly. When settled under satisfactory conditions, people do not tend to migrate unless forced to do so. Nevertheless, there can be many reasons why a person would choose to migrate, and migration is more likely to occur at points in a person's life when disruption is to be expected, such as when leaving home, starting a job, or even retiring (Beaujot and Kerr, 2004, p. 104). One way to understand the decision to migrate is to identify **push factors**, specific unfavourable conditions that encourage an individual or group to leave a place, and **pull factors**, local attractive conditions that may cause them to choose a particular destination. For instance, poor economic conditions, political persecution, warfare, or localized health threats (such as malaria or cholera) may serve as motivation to leave, while better economic conditions, political freedom, or even a common language may prove attractive. In practice, these push and pull factors do not necessarily work independently, and, as a result, these factors may exist as an "interplay of political, social, economic, legal, historical and educational" conditions in both the destination and the source communities (Mejia et al., 1979, p. 102).

In this section you will conduct an "interview" with either an imaginary migrant or a person whom you may know who has made a long-distance, permanent move between countries, provinces or major regions in the past. Alternatively, you may interview a classmate or answer the questions yourself based on your own experiences. The focus of your interview will be on the motivations for carrying out the migration, both in the reasons for leaving and the reasons for choosing the particular destination. Your object, then, is to examine and illuminate the push and pull factors that directed this person's (real or imagined) migration. You can set the context, including the date and places of origin and destination, however you prefer, but the stated factors should fit that context.

Once you have completed your interview fill in the table below. Provide sufficient detail to understand how these factors affected the decision to migrate.

Place and date of Origin	
Destination	
Push Factors for Migration	

Pull Factors for Migration	

References

Alberta. 2009. http://www.finance.alberta.ca/aboutalberta/population_reports/2009_1stquarter.pdf.

Alberta 2012. http://www.finance.alberta.ca/aboutalberta/population_reports/2012-4thquarter.pdf.

Beaujot, R.P., & Kerr, D. 2004. *Population change in Canada.* Don Mills, ON: Oxford University Press.

Encyclopedia of Saskatchewan. 2006. esask.uregina.ca/entry/population_trends.html.

Mejia, A. et al. *Physician and Nurse Migration: Analysis and Policy Implications.* Geneva: World Health Organization, 1979.

Statistics Canada. 2009a. http://www.statcan.gc.ca/estat/estat-eng.htm.

Statistics Canada. 2009b. http://www.statcan.gc.ca/edu/power-pouvoir/ch12/5214890-eng.htm.

Trovato, Frank. *Canada's Population in a Global Context.* Don Mills, ON: Oxford University Press, 2009.

US Census Bureau. 2009. http://sasweb.ssd.census.gov/idb/ranks.html.

Module 6
Cultural Landscapes

Introduction

Health and culture are fundamentally linked, and what we do and how we do it has much to say about our patterns of health, as does our culturally mediated reaction to disease (Trostle, 2004, p. 6). This is especially the case for indigenous societies around the globe, whose behaviour and beliefs may leave them exposed to a heavier burden of disease than western cultures. According to Kunitz (1996, p. 177), "diseases rarely act as independent forces but are shaped by the contexts in which they occur; . . . [and] the health consequences of social change are mediated by the way people are incorporated into national and international economies, as well as by their own cultural values and social organization. . . ."

In this module you will explore the relationship between culture and health, paying particular attention to the case of Canada's Aboriginal peoples. To complete this module you will examine the changes in cause of death that occurred among the Métis people during the last century. As well, you will participate in a group discussion about the causes and control of an epidemic disease that is currently affecting many First Nations communities in Canada.

Finally, you will carry out two exercises that examine culture and identity. In the first of these you will answer ten quick questions about the nature of Aboriginal languages in Canada based on a brief reading. In the second of these you will be asked to identify what, if anything, makes us identify as Canadians.

Key Concepts

epidemiologic transition	mortality	diabetes mellitus
Métis health	First Nations health	culture change and health
language	identity	

Learning Objectives and Skill Development

1. To apply Omran's theory of the epidemiologic transition to the mortality patterns of the Métis people.
2. To appraise the role of culture change in affecting Aboriginal mortality patterns over time.
3. To assess and explain the causes of the current diabetes epidemic among the First Nations of Canada.
4. To identify and describe the Aboriginal languages of Canada.
5. To speculate as to the components of the Canadian identity.

Tools Required

Internet access

Student Name: _____

Student Number: _____

Course/Section: _____

A. The Métis and the Epidemiologic Transition

The diseases that kill Canadians today are, for the most part, not the same ones that killed our ancestors a century ago. Then, infectious diseases such as tuberculosis, diphtheria, and influenza were leading causes of death. Now, those diseases appear far down the list, and they enter the top 10 only when aggregated as a whole (see Table 1). Instead, they have been replaced in importance by diseases of the circulatory system, especially heart disease and strokes, by the various cancers, and a few others, including diabetes. Some of these are considered lifestyle diseases, as their roots lie in the fundamental changes that we have made to our way of life in the modern era. Others, including cancer, are deemed diseases of old age, only becoming common as our lifespans have increased, since in most cases they appear as we enter middle age and beyond.

Table 1: Leading Causes of Death in Canada, All Ages, 2005

Causes	# of deaths	Deaths per 100,000 pop.
Diseases of the circulatory system	71,749	222.5
Cancer	68,790	213.3
Diseases of the respiratory system	20,484	63.5
Diseases of the nervous system	10,796	33.5
Endocrine, nutritional, & metabolic disease	10,266	31.8
Unintentional injuries	9,505	29.5
Diseases of the digestive system	8,952	27.8
Mental disorders	7,750	24
Genitourinary diseases	5,200	16.1
Infectious & parasitic diseases	4,154	12.9

This pattern of change in mortality was first expressed in broader form by Abdel Omran (1971). His theory of the **epidemiologic transition** suggested that in a given society improvements in nutrition, rising standards of living, advances in medical capacity, and increases in longevity, can spark a long-term shift in patterns of mortality. Omran believed that this transition went through three phases: an **age of pestilence and famine**, marked by high mortality, especially infant mortality, poor life expectancy, and the dominance of infectious diseases; an **age of receding pandemics**, marked by diminishing mortality (especially among children), increased life expectancy, and declining importance of epidemics; and an **age of degenerative and man-made diseases**, in which mortality continues to decline in all age groups, life expectancy increases, and the dominant causes of death are of degenerative and man-made diseases.

Although the determinants of this transition were said to be complex, Omran (1971, p. 520) argued that cultural changes play a key role. Thus, he singled out "socioeconomic, political and cultural determinants includ[ing] standards of living, health habits and hygiene and nutrition" as major factors. Omran's theory has been applied in many contexts and has proved useful as a descriptive tool. However, Young (1988) concluded that transition theory could be applied to the history of Subarctic First Nations with only mixed results. Though he could broadly distinguish the three eras, the experience of these people did not seem to fit Omran's proposed progression. While the degenerative and lifestyle diseases have begun to emerge, it would appear that the infectious diseases have not receded into the distance. Moreover, the Subarctic people continue to experience high rates of accidents and violence, far higher than would be predicted by Omran's model.

In this section you will consider the epidemiologic transition model while examining historical and contemporary data for cause of death among another Aboriginal group in Canada, the Métis. On pages 68 to 69 you will find a table listing causes of death of those who died between 1928 and 1934 around the St Henri Mission in the Vermilion district of northern Alberta. Although ethnicity is not noted, the majority of those listed would have been Métis. At this time the Métis of this region lived a traditional lifestyle, which would have centred on hunting, trapping, and fishing. As a result, they tended to be highly mobile, and their vigorous activities would have required a substantial expenditure of energy, especially during hunting and trapping seasons. Their foodstuffs were almost entirely local and taken from the bush, and so game animals, wild birds, and fish featured heavily in their diet, as did berries and other gathered foods. At this time non-native foods would have been very expensive, and the Métis local diet would have been supplemented only by small amounts of flour, baking powder, lard, sugar, salt, beans, jam or syrup, and bacon.

The entries are as recorded in the log, though some spelling errors have been corrected. Examine Table 2 (pp. 68–9) and answer the following questions. In order to do so you will have to do a bit of detective work because some of the terms used are no longer in common use.

1. What was the average age of those who died of tuberculosis?

2. What was the average age at death overall?

3. Indicate in the table what percentage of people died of the following causes:

Causes	Percentage
a. Infectious disease	
b. Accident, homicide, or poisoning	
c. Cancer	
d. Heart disease or stroke	
e. Kidney disease	
f. Other	

4. Which stage of the epidemiologic transition would best describe these people?

5. The people of the St Henri Mission were comparatively isolated from the major urban centres of the south. What differences in cause of death might you expect to find among those living in the city of Edmonton at that time?

6. Compare the causes of Métis death from northern Alberta for the period 1928–34 to those for Canadian Métis in general for the period 1991–2001, found in the Statistics Canada publication, "Mortality of Métis and Registered Indian Adults in Canada: An 11-year Follow-up Study" (Statistics Canada, 2009). By the later period Métis culture had changed radically. How had this affected the major causes of death?

7. In your opinion, does the experience of the Métis people tend to support Omran's epidemiologic theory? Explain.

Table 2: Causes of Death, St Henri Mission, Alberta, 1928–1934

Sex	Place of Death	Age	Cause of Death
M	Fort Vermilion	70	Enlarged prostate
F	Fort Vermilion	2	Tuberculosis of bowels
M	Little Red River	7	Tuberculosis
M	Little Red River	3	Tuberculosis
F	Little Red River	6 Months	Tuberculosis
M	On ice of the Peace R.	35	Freezing
F	Fort Vermilion	13	Pulmonary Tuberculosis
M	Fort Vermilion	7	Pulmonary Tuberculosis
F	Fort Vermilion	15	Septicaemia
F	Fort Vermilion	15	Tuberculosis
M	Two Lakes, Ft Verm.	56	Kidney Stone
F	Little Red River	61	Paralysis
F	North Vermilion	10	Tuberculosis
F	Carcajou	75	Pulmonary Tuberculosis
F	Keg River Prairie	16	Tuberculosis
M	Two Lakes	58	Heart lesion
F	Fort Vermilion	13	Pulmonary Tuberculosis
F	Fort Vermilion	9	Pulmonary Tuberculosis
M	Keg River Prairie	3	Tuberculosis
F	Keg River Prairie	2	Tuberculosis
M	Tall Cree Prairie	2 Days	Premature death
M	Hay River	7 Days	Premature death
F	Carcajou Point	10 Months	Whooping cough
F	Fort Vermilion	18	Tuberculosis
M	Fort Vermilion I.R.	1 ½	Lobar pneumonia
M	Fort Vermilion Hosp.	19	Acute pneumonia
F	Fort Vermilion Buffalo Prairie	16 Days	Probably general debility
M	Carcajou Point	5	Pulmonary Tuberculosis
M	Carcajou Point	83	Old age
F	Keg River Prairie	9 Months	Poison (herb)

F	Keg River Prairie	18	Pulmonary Tuberculosis
M	Keg River Prairie	6 ½	Pulmonary Tuberculosis
M	Fort Vermilion	12	Pulmonary Tuberculosis
M	Eleske	6	Tuberculosis of bowel
F	Fort Vermilion	48	Pulmonary Tuberculosis
M	Fort Vermilion	79	Chronic Nephritis
M	I.R. 164	11	Tuberculosis
M	Fort Vermilion	Stillborn	Stillborn
M	Fort Vermilion	10	Tuberculosis meningitis
M	I.R. 164A	2 ½	Pulmonary Tuberculosis
M	Fort Vermilion	61	Cardio decomposition
M	North Vermilion	19	Pulmonary Tuberculosis
M	Fort Vermilion	23	Pulmonary Tuberculosis
F	Fort Vermilion Hosp.	8 Months	Dysentery
F	Fort Vermilion Hosp.	50	Tuberculosis and meningitis
F	Fort Vermilion	13	Tuberculosis
M	Hay River	22	Tuberculosis
M	Keg River Prairie	4	Grippe (influenza)
M	Keg River Prairie	2	Grippe (influenza)
F	Little Red River	11 ½	Pneumonia
M	Apakwatissipik	79	Physical Exhaustion
F	Fort Vermilion Hosp.	9	Erysipelas
F	Hay Lake (Revill Post)	2 ½	Pulmonary Tuberculosis
M	Hay Lake	23	Tuberculosis
M	Fort Vermilion	58	Tuberculosis Peritonitis
F	Little Red River	7	Pulmonary Tuberculosis
F	Paddle River Prairie	2	Pulmonary Tuberculosis
M	Fort Vermilion Hosp.	4 ½	Tuberculosis Meningitis
F	I.R. 164A	16	Tuberculosis

Source: Provincial Archives of Alberta (1928–34).

B. Type 2 Diabetes Mellitus and Canada's First Nations: The Role of Culture

In this section you will be asked to participate in a group discussion, either with the entire class or a smaller group (your instructor will determine this). First, read the case study below. Then consider the questions that follow the case study. As you read, you can make point-form notes in the space provided, so that you can contribute to the discussion.

In addition, you may wish to refer to the following resources in order to learn more about this subject:

- Young, T.K., Reading, J., Elias, B., and O'Neil, J.D. 2000. Type 2 diabetes mellitus in Canada's First Nations: Status of an epidemic in progress. *Canadian Medical Association Journal, 163*(5), 561–6. http://www.cmaj.ca/cgi/content/full/163/5/561.
- Oster, R.T., Virani, S., Strong, D., Shade, S, and Toth, E.L. 2009. Diabetes care and health status of First Nations individuals with type 2 diabetes in Alberta. *Canadian Family Physician, 55*(4), 386–93. http://www.cfp.ca/cgi/content/full/55/4/386.
- Dyck, R., Osgood, N., Lin, T.H., Gao, A., and Stang, M.R. 2010. Epidemiology of diabetes mellitus among First Nations and non-First Nations adults. *Canadian Medical Association Journal, 182*(3), 249–56. http://www.cmaj.ca/cgi/content/full/182/3/249.
- Munro, M. First Nations diabetes strategy is contentious. *Canwest News Service*, Canada.com. http://www.canada.com/health/First+Nations+diabetes+strategy+contentious/2082427/story.html. Accessed May 19, 2010.

Case Study

Type 2 diabetes mellitus (T2DM) is a disease of the endocrine system in which the body is unable to process and eliminate the glucose that is naturally found in the blood, due to insulin resistance. Although diabetes can often be controlled without medication, usually by making lifestyle modifications involving diet and exercise, those with uncontrolled diabetes face a heightened risk of severe complications as blood sugars build up and attack the peripheral vascular system. These complications may include diabetic retinopathy leading to blindness, coronary artery disease, kidney problems, amputations, and, in some cases, death. Currently, more than 3 million Canadians have diabetes, and this number may reach 3.7 million within a decade (Canadian Diabetes Association).

The list of risk factors for T2DM at the individual level is extensive. Excess weight, especially obesity, and lack of exercise (a sedentary lifestyle) are perhaps the main factors. However, diet, as it relates to increased weight and high blood pressure and cholesterol levels, is also thought to play a significant role. Even education and income can also contribute indirectly, because those with less formal education and lower income are more likely to smoke, to have poorer diets, and to be more sedentary. Other risk factors are advanced age, family history of T2DM and having had gestational diabetes. (Gestational diabetes is a temporary condition in which pregnant women suffer from high blood-glucose levels. Being born to a mother with gestational diabetes also increases risk.) Finally, those belonging to certain ethnic populations are at much greater risk, including, for example, those of Aboriginal, African, Hispanic, and Asian descent. Among the most vulnerable in Canada are the First Nations people.

Although the prevalence of diabetes continues to increase among the general Canadian population, among the First Nations, the disease has become an epidemic. It is difficult to determine how many have T2DM because many are unaware that they have it and thus go unrecorded in the statistics; however, it is estimated that the age-adjusted prevalence rate for First Nations men is 3.6 times that of Canadian men in general, and that the rate for First Nations women is 5.3 times that of Canadian women in general (Young et al., 2000). In Saskatchewan the rates are 2.5 (male) and 4 (female) times the non-First Nations rates, and it is estimated that 50% of all First Nations women aged 60 and over and more than 40% of the men have diabetes (Dyck et al., 2010). In Manitoba, the combined rate is 4.5 times higher for First Nations people (Green et al., 2003).

In addition, the age of onset among First Nations people would seem to be much younger than among the broader population, and it is decreasing. In the past, T2DM was called adult-onset diabetes because it was generally a disease that was experienced later in life. This has changed, and young Aboriginal children are now being diagnosed with a disease that in the past would have been limited to seniors. Finally, there is some indication that those who live traditional lives may gain some protective benefit against T2DM, for in some provinces the more southern First Nations, who live less traditional lives, tend to have higher rates of diabetes (Green et al. 2003). Lower rates of T2DM among those leading more traditional lives may be due to their greater consumption of traditional foods, their more active lifestyles, or both.

This epidemic is of recent origin among the First Nations people. There is little evidence of diabetes among them before the Second World War; only in the 1980s did physicians begin to notice a troubling increase in the incidence of T2DM. There is no single, clear-cut reason for this epidemic. Indeed, different researchers have tended to focus on substantially different explanations. With this difference in approach and emphasis have come differences of opinion on how to combat the disease. The major areas include:

a. **Lifestyle:** The increasing trend among First Nations people, especially children, towards leading a sedentary lifestyle and eating unwholesome foods has been cause for much concern. These habits may be tied to, for instance, increased television watching, decreased exercise levels, and a tendency to prefer highly processed, western foods. In this view, the focus is on harmful behaviour that, hopefully, can be addressed through education and medical intervention with the individual.

b. **Culture:** Others acknowledge the role of such individual behaviour, but point to broader, damaging changes in First Nations societies that have created a cultural environment favouring the emergence of T2DM. Dr Kue Young, a pioneering researcher on the subject, concluded that "Diabetes can be considered to be indicative of the rapid sociocultural changes experienced by Aboriginal people in the past several decades" (Young, 1994). Some researchers argue that these changes must be seen within a colonial framework, and would focus more on the influence of civilization—in particular, the attempt by the government and churches to force Aboriginal people to adopt western culture. From this perspective, residential schools, loss of language and the traditional economy, inadequate housing, lack of income and employment, and adoption of western foods, are all to be taken into account. The implication of this approach is that Canadian society, and not simply the individual, is in large part to blame for the epidemic of T2DM. Furthermore, any attempt at a solution must acknowledge the broader picture: "The design of successful education and prevention programs as well as the delivery of optimum health care is dependent on understanding the client's frame of reference. The frame of reference includes cultural beliefs and attitudes that impact directly on the health and behaviour of the client" (Joe and Young, 1994, p. 11).

c. **Genetics:** In the past, researchers into T2DM have theorized that Aboriginal people in North America have been prone to obesity, and thus diabetes, because of a genetic adaptation known as the "thrifty gene." It has been suggested that this hypothetical gene or gene combination enabled Aboriginal people in the past to store energy more efficiently in the form of fat for longer periods of time in an environment where food was scarce. This is contrasted with people who are descended from populations—such as those from Europe—that had long since become adapted to the stable food supply provided by agriculture. In the modern era, with food as close as the corner store, and with a minimal income assured, the result has been a tendency for Aboriginal people to overeat and to accumulate fat, leading to high rates of obesity. More recent research has identified genetic variants in some communities that are now known to lower the age of onset of T2DM, and that may increase the risk of developing the disease. However, even in this case, researchers acknowledge the major role that the environment, through modification of human behaviours, has played in the emergence of this epidemic (Hegele and Bartlett, 2003). Nevertheless, the focus on genetic factors, with its implied accusation of an inherent weakness in Aboriginal people, combined with a lack of treatment options, may influence health policy makers to limit their interventions.

d. **Physical Environment:** Most recently, new research has emerged linking diabetes to environmental pollution. Some studies point to ties between T2DM and contaminants, including pesticides, persistent organic pollutants (POPs), and PCBs. This possible link might perhaps explain the emergence of the epidemic in First Nations communities (Everett and Matheson, 2010; Carpenter, 2008).

Questions for Discussion

1. What is the underlying cause of the current First Nations diabetes epidemic?

2. What approach(es) would you take towards combatting this epidemic?

3. Would a single approach work for southern and more isolated communities? If not, how might you modify your approach?

4. Are there lessons to be learned from the First Nations situation regarding the growing obesity crisis in the broader Canadian population? What are they?

5. How might health-care providers incorporate ideas from traditional First Nations life and culture into treatment programs? In your opinion, would this be likely to increase the effectiveness of these programs?

6. In your opinion, which of the main risk factors for diabetes offers the greatest possibility for intervention at the community level?

7. Should the cost of transporting healthy foods to remote communities be fully subsidized by the federal government?

8. Should legal or regulatory measures be taken against foods that are known to cause obesity and contribute to the diabetes epidemic (e.g., banning fast foods, trans fats or high-fructose corn syrup; imposing tariffs or special taxes to discourage consumption of such foods and offset treatment costs; and penalizing parents who give unhealthy food to their children, etc.)?

C. Aboriginal Languages in Canada

Language constitutes a fundamental component of culture, serving multiple roles. It serves to transmit knowledge between generations and to maintain social order, to reinforce connections between people with shared traits, and also to differentiate between those whose cultures differ. Although the broader focus on issues of language and culture in Canada surrounds the two official languages, English and French, our linguistic landscape is actually much richer. Indeed, the most recent Canadian census recognized over sixty Aboriginal languages in use in the country, with almost 213,500 people reporting an Aboriginal first language learned at home. Not all were equal in either the number of speakers, or their geographical extent, however. Some are spoken by only a few tens of people in a relatively small territory, while others include tens of thousands of speakers across Canada.

For this section you will be answering a series of questions concerning Aboriginal languages in Canada. Begin by reading the report "Aboriginal languages in Canada," which can be found on the Statistics Canada site (http://www12.statcan.gc.ca/census-recensement/2011/as-sa/98-314-x/98-314-x2011003_3-eng.pdf). Once you've finished that, answer the questions in the table below. Each answer requires only a single word or phrase.

	Question	Answer	
1	The most commonly spoken Aboriginal language in Canada is _____.		
2	_____ is the traditional language of the Métis people.		
3	Most Inuit speakers are found in _____.		

4	The highest proportion of Aboriginal speakers (compared to the total population) is found in which province?		
5	The lowest proportion of Aboriginal speakers to total population is found in _____.		
6	The Blackfoot language is mainly spoken in _____.		
7	The Dene language is classified as part of which linguistic family?		
8	Which language group is most common in Manitoba?		
9	The Aboriginal language with the smallest reported number of speakers in Canada is _____.		
10	The province with the largest number of Aboriginal languages spoken is _____.		

D. Identity in Canada

It is said by some that there is no Canadian culture. Rather, they argue, we have little in common as a nation, perhaps excepting a shared history of sorts. While there may be regional cultures, or broader ethnic cultures, it is sometimes said that nothing serves to tie us together. Others argue that there is a Canadian culture, and that we can point to certain cultural attributes that provide an underlying framework of identity. In this exercise you will list five things that may be used to define us as Canadians, and argue for or against their role in creating a shared identity.

References

Canadian Diabetes Association. http://www.diabetes.ca/about-diabetes/what/prevalence. Accessed May 19, 2010.

Carpenter, D.O. 2008. Environmental contaminants as risk factors for developing diabetes. *Review of Environmental Health, 23*(1), 59–74.

Dyck, R., Osgood, N., Lin, T.H., Gao, A., & Stang, M.R. 2010. Epidemiology of diabetes mellitus among First Nations and non-First Nations adults. *Canadian Medical Association Journal, 182*(3).

Everett, C.J., & Matheson, E.M. 2010. Biomarkers of pesticide exposure and diabetes in the 1999–2004 National Health and Nutrition Examination Survey. *Environment International 36*(4), 398–401.

Green, C., Blanchard, J.F., Young, T.K., & Griffith, J. 2003. The epidemiology of diabetes in the Manitoba-registered First Nation population: Current patterns and comparative trends. *Diabetes Care, 26*(7).

Joe, J.R., & Young, R.S. 1994. Introduction. In J.R. Joe and R.S. Young, *Diabetes as a disease of civilization: The impact of culture change on indigenous peoples* (pp. 1–18). Berlin: Mouton de Gruyter.

Hegele, R.A., & Lloyd, C.B. 2003. Genetics, environment and type 2 diabetes in the Oji-Cree population of Northern Ontario. *Canadian Journal of Diabetes, 27*(3), 256–61.

Kunitz, S.J. 1996. *Disease and social diversity: The European impact on the health of non-Europeans.* New York: Oxford University Press.

Omran, A.R. 1971. The epidemiological transition: A theory of the epidemiology of population change. *Milbank Memorial Fund Quarterly, 49,* 509–38.

Provincial Archives of Alberta. 1928–34. No. 87.385. Box 29 Items 850; 850A Community and Occupational Health, Department of Vital Statistics Branch, Register of Baptisms, Marriages and Burials Recorded by Church Ministers and Civil Authorities throughout Alberta, 1928–1931; 1932–1934.

Public Health Agency of Canada. 2008. http://www.phac-aspc.gc.ca/publicat/lcd-pcd97/pdf/lcd-pcd-t1-eng.pdf. Accessed May 19, 2010.

Public Health Agency of Canada. 2009. http://www.phac-aspc.gc.ca/cd-mc/diabetes-diabete/index-eng.php. Accessed May 19, 2010.

Statistics Canada. 2009. http://www.statcan.gc.ca/pub/82-003-x/2009004/article/11034/tables/tbld-eng.htm.

Statistics Canada. 2011. http://www12.statcan.gc.ca/census-recensement/2011/as-sa/98-314-x/98-314-x2011003_3-eng.pdf. Aboriginal languages in Canada, Language, 2011 Census of Population. Accessed May 12, 2013.

Trostle, J.A. 2005. *Epidemiology and culture,* Cambridge Studies in Medical Anthropology 13. Cambridge, UK: Cambridge University Press.

Young, T.K. 1988. Are Subarctic Indians undergoing the epidemiologic transition? *Social Science & Medicine, 26*(6), 659–71.

Young, T.K. 1994. *The health of Native Americans: Towards a biocultural epidemiology.* Oxford: Oxford University Press.

Young, T. Kue, Reading, J., Elias, B., & O'Neil, J.D. 2000. Type 2 diabetes mellitus in Canada's First Nations: Status of an epidemic in progress. *Canadian Medical Association Journal, 163*(5), 561.

Module 7
Social Well-Being

Introduction

Though poorly defined, the concept of **social well-being** has been an integral part of our understanding of health since the World Health Organization's declaration in 1948 that "Health is a state of complete physical, mental and social well-being and not merely the absence of disease or infirmity" (WHO, 1948). If we wish to narrow it down, we might say that social well-being is a situation in which we have access to the economic, health, and educational resources we need to function at a minimally satisfactory level.

Health and social well-being are well understood to vary across space and between populations, from the global down to the local scale. For instance, most of us will recognize that the patterns of health (and available resources) are different for those living in the developing world than for the citizens of the wealthiest countries. However, there may also be substantial spatial variations in well-being in your community, where some neighbourhoods may be much better off than others.

One example of this disparity relates to **food security**. Food security simply means consistent physical and economic access to an adequate and safe supply of nutritious foods that will support a healthy life. Even in Canada many people are subject to *food insecurity*. A geographical expression of food insecurity is the **food desert**. A food desert is a populated area, often within the core of a major city, where the residents have little or no access to reasonably priced healthy foods. Usually this means that there is no large grocery store in the vicinity and that food is bought instead at either small corner stores or businesses whose primary function is something other than selling food (e.g., drugstores or convenience stores).

These deserts emerged when two related trends began to unfold in North American cities following the Second World War (Larsen and Gilliland, 2008). First, this era saw the rise of the automobile culture, with increased mobility for the more affluent, and a growing view of cities as unhealthy places. This encouraged a mass movement of wealthier people into the suburbs and, in some cities, led to the economic decline of older neighbourhoods where only the poor remained. Secondly, the emergence of chains of retail grocery corporations, combined with an increase in the physical size of individual stores (up to the size of the modern "superstore"), prompted a move to outlying areas where large amounts of cheap land were available for parking and floor space. Together, these trends encouraged food sellers to follow their wealthier customers to the suburbs and to abandon the core areas.

For the most part, people living in food deserts will not find enough vegetables, fruits, and other nutritious foods in local stores. Moreover, research has shown that even when such healthy foods are available, the poor actually pay more than those who have access to supermarkets in the more affluent suburbs because local businesses tend to face competition and must adjust their prices accordingly. Note, however, that there are exceptions. Researchers have found that food deserts have not emerged in every city, and in some cases poorer areas have been found to have superior access to grocery stores than more economically advantaged areas in the same city. Nevertheless, studies have also identified deserts in many Canadian cities.

In this module you will explore differences in access to affordable healthy foods in an urban setting and the consequences of those differences. You will map major grocery stores within a neighbourhood, determine the availability and cost of selected foods at two types of stores, and create a basic economic profile of that neighbourhood. Finally, after reading a report on obesity in Canada, you will identify some of the specific health outcomes that are associated with a poor diet. To complete the three parts of this module you will draft a map, conduct a simple field survey, consult the Canadian census, and answer several questions based upon a research report.

Key Concepts

food desert   Healthy Food Basket   food insecurity
obesity

Learning Objectives and Skill Development

1. To assess a study area for the presence of a food desert and evaluate the potential impact of such a desert on people of limited economic means.
2. To determine the differences in the availability, cost, and quality of food between grocery stores and convenience stores.
3. To describe some of the health implications of inadequate access to healthy foods.

Tools Required

Internet access

Student Name: _____

Student Number: _____

Course/Section: _____

A. Food Deserts

In this section you will map grocery stores in your study area in order to determine if this area contains a food desert. You will also visit two different types of food retailer in order to compare the variety and quality of foods available as well as their relative prices.

Mapping for Food Deserts

Begin by either selecting a neighbourhood for your study or reviewing your assigned area (your course or lab instructor will explain which of these options you are to choose). These neighbourhoods will be based on census tracts (CTs) for the purposes of this exercise. However, please note that census tracts are not used in either Prince Edward Island or the Canadian territories. For these places, you will have to use alternative enumeration areas or neighbourhoods.

A census map for your study area can be found online at the Statistics Canada site at http://geodepot.statcan.gc.ca/2006/13011619/03200401_05-eng.jsp. Simply input the name of your province, Census Metropolitan Area (the CMA, or city), and census tract in the three pull-down menus, being sure to click the **next** button between the first two. Then you will click the **download** button. Once you have loaded the map into your browser, you can either print or save the file, depending on how your instructor prefers you to process the map. If your community is not organized into census tracts—this would include smaller communities, as well as the places mentioned above—you can locate and download maps by town or other area from another Statistics Canada web page: http://geodepot.statcan.ca/Diss/Maps/ReferenceMaps.

In the first step you will find the locations of major grocery stores within your study area. We will define a grocery store as a business whose primary function is to sell a wide variety of healthy food items for the purposes of further processing and consumption in the home. You do not have to visit the study area in order to locate these businesses, though. Instead, you can use your local Yellow Pages and Internet resources (e.g., Google Maps, Internet listings, etc.). Once you have identified all of the grocery stores in your area, plot them on your map using a clear, legible symbol. You do not need to name them. Do remember to put your name, student number, class number, date, a title and legend, and any other information that your instructor may request on the map.

Once you have mapped the locations, you will want to identify the food deserts. You will do this by drawing circles representing 1 km buffers (i.e., with a radius of 500 metres) around each location. This is considered to be equivalent to a reasonable walking distance. Once the map is complete, answer the questions on the following pages.

1. By referring to your map, what percentage of the study area would you classify as a food desert?

   _____

   _____

   _____

2. In our exercise we are simply looking at the extent of the study area outside of the 1-km buffers and factoring in walking distance when considering whether there is a food desert or not. How would your analysis be different if you (1) had data for population density within the area, and (2) considered the availability of mass transit (such as bus or subway lines)?

   _____

   _____

   _____

   _____

   _____

   _____

   _____

3. Why might you want to include on your map the grocery stores in census areas that are right next to your study area?

   _____

   _____

   _____

   _____

   _____

Surveying Food Prices and Availability

The second part of this section examines the availability and costs of foods in two different types of business: (1) grocery stores or supermarkets, which we might find more commonly in the suburbs or on their margins; and (2) convenience stores, which can sometimes serve as primary food stores in core areas. Please note that you do not have to visit a store in your study area. Instead, you can visit a representative establishment in another part of your community, if you wish.

The table on the next page is derived from Canada's Healthy Food Basket, a selection of healthy foods that is used to measure food security in different neighbourhoods or regions (Tasnim and Shoveller, 2003). Your task here is to visit both types of food retailers and determine if the food items are available and, if so, what they cost and what their relative quality is. Remember to assess equivalent products and amounts. You will then answer the following questions.

1. From your analysis, how do the grocery store and the convenience store compare in terms of the foods they sell?

2. In your opinion, what are the disadvantages of living in an area that is served only by convenience stores or similar food retailers?

3. If the healthy foods are unavailable, too expensive, or of poor quality, what alternative foods might be available?

Food Item	Grocery Store or Supermarket			Convenience Store		
	Cost	Availability	Quality	Cost	Availability	Quality
Iceberg (head) lettuce, 450 g						
2% milk, 4 l						
Eggs, Grade A large, 12						
Corn Flakes®, 675 g						
Hot dog/hamburger buns, 8 pack						
Cabbage, 255 g						
Canned pink salmon, 213 g						
Regular ground beef, 655 g						
Processed cheddar cheese slices, 500 g						
Carrots, 1.1 kg bag						
White beans, dry, 454 g						
Medium cheddar cheese, 227 g						
Long grain white rice, parboiled, 900 g						
Macaroni and cheese dinner, 225 g						
Butter, salted, 454 g						
Canned corn, 341 ml						
Oranges, 710 g						
Canned baked beans in tomato sauce, 398 ml						
Peanut butter, 500 g						
Frozen orange juice concentrate, 355 ml						
Canned fruit cocktail, 398 ml						

B. Income and Food Deserts

It would seem that the effect of living in a food desert varies with economic status. Individuals or families with higher incomes may experience few or even no adverse consequences from the limited local supply of healthy food because they can afford to travel longer distances outside of their neighbourhoods for better-quality food, or to obtain more expensive, but healthier, alternatives closer to home (such as having fresh produce delivered directly to their door or buying more expensive vegetables at a farmers' market). Conversely, lower income families may be limited to the poorer-quality local choices owing to limited transportation options.

In this section you will explore the economic profile of your study area as defined by family status, income, education, and employment (see table below). We will use these values as indicators, or proxies, for poverty. A proxy is a measurable variable used in place of a related quality that is not measurable. You will also compare these values to the average values for your town or city in order to assess whether your study area is better or worse off, in general, than the larger community. Finally, by referring to the figures in your table, you will answer a few questions. Complete the table below. Begin by obtaining the census data for your study area at the following Statistics Canada site: http://www12.statcan.ca/census-recensement/2006/dp-pd/prof/92-597/index.cfm?Lang=E

You will need to use one of these three methods to access the data: entering a postal code, using GeoSearch2006, or entering a CMA/CA code and census tract name. Your course or lab instructor will decide which method is suitable in your case. Once you have accessed the census tract profile for your study area, find the relevant information and enter it in the table. Note that you may have to calculate the value for some indicators from more than one census category. Once again, note that since Prince Edward Island and the Canadian territories, as well as smaller communities in other provinces, are not divided into census tracts, so you may need to find your data from a Statistics Canada website that reports data for individual communities: http://www12.statcan.ca/census-recensement/2006/dp-pd/prof/92-591/index.cfm?Lang=E

Your instructor can help you decide which information you will require.

Economic Indicators for Study Area, 2006 Census of Canada

Indicator	Study Area (CT, CA, or Town)	Town/City or Province
1. Percentage of families that are female lone-parent families		
2. Median income in 2005—All census families ($)		
3. Percentage of population 15 years and over with a university certificate, diploma, or degree		
4. Unemployment rate		

1. How are the proxies that you are measuring related to poverty, and how might they affect access to healthy food?

2. Referring to the values in your completed table, describe how the study area compares to your town or city as a whole. For those working with a census agglomeration or other unit, compare your data to the provincial figures.

3. Judging by the location of grocery stores in your study area and the economic profile, is it likely that this population is at risk from living in a food desert? Why or why not?

C. Health Consequences of Food Insecurity

One of the main health problems associated with a poor diet is obesity. In this case, a "poor" diet might be lacking, not in calories, but in nutrients. Although the relationship between socio-economic status, genetics, behaviour, the built environment, and obesity is complex, studies have found that, among adults in Canada, obesity is associated with low consumption of fruits and vegetables combined with excess consumption of highly processed, sugary, or high-fat foods, especially foods rich in saturated fats. These are characteristics that we might expect to see among people of limited resources living in a food desert—people who cannot travel longer distances for healthier alternatives.

Over the past three decades adult obesity rates in Canada have increased substantially, rising from 14% of the adult population in the late 1970s to 23% in 2004, and another 36% of adult Canadians are now classified as overweight (Tjepkema, 2005). This crisis is not unique to Canada, but is part of an emerging global epidemic that has seen more than 1 billion overweight adults and 300 million or more clinically obese (WHO, 2010). Ironically, in the developing world this epidemic is tied to forces that we might otherwise see as positive, including economic development, modernization, and increasing urbanization.

In this section you will answer several questions about obesity and its effects on human health and well-being. Please refer to the online document "Adult Obesity in Canada: Measured Height and Weight" by Michael Tjepkema (http://www.statcan.gc.ca/pub/82-620-m/2005001/pdf/4224906-eng.pdf) when answering the following questions:

1. How do health researchers define obesity? What are the three classes of obesity that they recognize?

2. What are the major health conditions associated with obesity?

3. In 2004 what age group(s) (a) had the highest rates of adult obesity in Canada, and (b) had the lowest?

4. In 2004 which two provinces had the highest percentages of adults classified as overweight or obese (male and female)? Which two had the lowest percentages? Include the values.

References

Larsen, K. & Gilliland, J. 2008. Mapping the evolution of 'food deserts' in a Canadian city: Supermarket accessibility in London, Ontario, 1961–2005. *International Journal of Health Geography*, 7(16). http://www.ij-healthgeographics.com/content/7/1/16.

Nathoo, T. & Shoveller, J. 2003. Do healthy food baskets assess food security? *Chronic Diseases in Canada*, 24 (2/3). http://www.phac-aspc.gc.ca/publicat/cdic-mcc/24-2/c_e.html.

Tjepkema, M. 2005. *Adult obesity in Canada: Measured height and weight.* Statistics Canada Cat. No. 82-620-MWE at 30. Ottawa: Statistics Canada. Available at http://www.statcan.gc.ca/pub/82-620-m/2005001/pdf/4224906-eng.pdf.

WHO. 1948. Preamble to the Constitution of the World Health Organization as adopted by the International Health Conference, New York, 19–22 June, 1946; signed on 22 July 1946 by the representatives of 61 States (Official Records of the World Health Organization, no. 2, p. 100) and entered into force on 7 April 1948.

WHO. 2010. http://www.who.int/dietphysicalactivity/publications/facts/obesity/en/.

Module 8
Political Geographies

Introduction

Political states play an important role in the global ordering of space, in shaping global processes, and in influencing human organization and behaviour within the state. The power and cohesiveness of a state is influenced by a variety of forces that might be at work. This includes forces that bind people and regions together and serve to strengthen the state, and forces that cause regional separation and weaken the state.

The influence of a state doesn't stop at its borders. States project power and influence globally in, for example, military affairs or more subtly in international trade and foreign aid. In some cases, the political decisions and actions in one state spill over and affect the political, social, and economic environment of neighbouring states. Consider, for example, the impacts on the lives of Canadians of tightened border policies and heightened border security between Canada and the United States post 9/11—particularly impacts on the everyday lives of those individuals who live near the Canada–United States border or who travel from Canada to the United States regularly.

In this module you will first examine the spatial political landscape in Canada based on results from the last three federal general elections. You will then examine participation in the Canadian political system and learn from your classmates about their voting behaviour. The module then focuses on borders and boundaries, exploring the permeability of borders and the changing Canada–United States border relationship. Finally, you will explore the politics of natural resources, specifically border disputes over natural resources and what constitutes "the national interest" with regard to political decisions about large-scale natural resource development.

Key Concepts

voting patterns
centripetal forces
territorial disputes
national interest

voting behaviour
political borders
positional disputes

centrifugal forces
border conflicts
functional disputes

Learning Objectives and Skill Development

1. To examine the political geography and voting patterns in Canada at the federal and provincial levels.
2. To examine spatial voting patterns for evidence of centrifugal or centripetal forces.
3. To explore the changing Canada–United States border and its influence on Canadians.
4. To explore border conflicts over natural resources.

Tools Required

Internet access

Student Name: _____

Student Number: _____

Course/Section: _____

A. Geography of the General Elections

Go to the Elections Canada website at http://www.elections.ca/. Follow the link to "Elections–Current & Past Elections," and then to "Past Elections." There you can find information on past general elections from 1997 to 2011, including maps of the official results of each general election.

1. Examine the map "Official Results of the 41st General Election" and briefly describe the geography of the election results. Repeat this exercise for the maps of the official results of the 40th and 39th general elections. Note any particular patterns you observe for each general election and across the last three general elections.

_____
_____
_____
_____
_____
_____
_____
_____
_____
_____
_____
_____
_____
_____
_____
_____
_____
_____

2. Centrifugal forces make it difficult to bind an area together as an effective state. Centripetal forces, in contrast, pull an area together as a single unit. On the basis of the geography you described in the previous question and referring to the general election maps for 2006, 2008, and 2011, can you identify any centrifugal or centripetal forces that might be at work in different political regions of the country? Explain what factors might be either separating political regions or binding political regions together. Consider, for example, potential economic, cultural or language differences amongst regions.

B. Who Participates?

Refer to the table on the following page, compiled from Elections Canada data for the last eight general elections. (You can view the data yourself by visiting the Elections Canada website at http://www.elections.ca/.) Then answer the questions that follow.

Voter Turnout in General Elections (%)

	1988	1993	2000	2004	1997	2006	2008	2011
NL	67.1	55.1	57.1	49.3	55.2	56.7	47.7	52.6
PE	84.9	73.2	72.7	70.8	72.8	73.2	69.0	73.3
NS	74.8	64.7	62.9	62.3	69.4	63.9	60.3	62.0
NB	75.9	69.6	67.7	62.8	73.4	69.2	62.9	66.2
PQ	75.2	77.1	64.1	60.5	73.3	63.9	61.7	62.9
ON	74.6	67.7	58.0	61.8	65.6	66.6	58.6	61.5
MB	74.7	68.7	62.3	56.7	63.2	62.3	56.1	59.4
SK	77.8	69.4	62.3	59.1	65.3	65.1	58.7	63.1
AB	75.0	65.2	60.2	58.9	58.5	61.9	52.4	55.8
BC	78.7	67.8	63.0	63.3	65.6	63.7	60.1	60.4
YT	78.4	70.4	63.5	61.8	69.8	66.1	63.2	66.2
NT	70.8	62.9	52.2	47.3	58.9	56.2	47.7	53.9
NU			54.1	43.9		54.1	47.4	45.7
Totals	75.3	69.6	64.1	60.9	67.0	64.7	58.8	61.1

Source: Elections Canada

1. Which province experienced the most significant increase in % voter turnout in 2011 compared to 2008?

2. Construct a bar graph below depicting the trend in % voter turnout for the general elections, 1988 to 2011, for i) your current province or territory of residence, and ii) the national total.

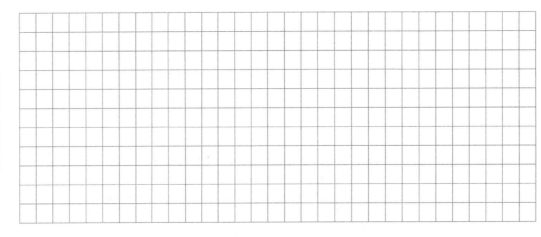

general elections, 1998–2011

3. Survey your classmates' voting behaviour and record your results in the table on the following page. Ask as many of your classmates as you can whether they voted in the most recent election (national, provincial, municipal, or other if non-Canadian). Tally the number of "yes" responses and the number of "no" responses. Record these numbers below and convert them into a percentage.

   _____ "Yes" ____ (%)   _____ "No" ____ (%)

4. For the "no" respondents, ask for the reasons that he or she did not vote. (a) What were the three most common reasons given by non-voters? (b) Would you speculate that these reasons reflect the Canadian national trend? Why or why not?

   _____
   _____
   _____
   _____
   _____

5. A report by Elections Canada (2008) indicates that in the 2006 general election, voter turnout for individuals 18–24 years of age was 43.8%; for 25–34 years, 49.8%; for 35–44 years, 61.6%; for 45–54 years, 70.0%; for 55–64 years, 75.4%; and for 65–74 years, 77.5%.

   a. Are the results of this report consistent with your survey results of your classmates?

      _____
      _____

   b. Voter turnout tends to increase beyond the 35-year age class. Can you explain why?

      _____
      _____
      _____
      _____

   c. What are some possible strategies that you would recommend for increasing voter turnout for the under 35-year age class in Canada?

      _____
      _____
      _____
      _____

Questions	"Yes"	"No"	Reasons for not voting
Did you vote in the last federal, provincial, or other election in your country, province, or city or state?			
Do you plan to vote in the next federal, provincial, or other election in your country, province, or city or state?			
Total (%)			

C. Tightening Borders

No land on the earth's surface remains "unclaimed," and no state exists in isolation of the political decisions of its neighbouring states. When one state makes political decisions in its own interests, the borders between that state and neighbouring states become more or less permeable. That is to say, while some borders may seem to persist on a map, many borders are quite fluid and their permeability changes as different national events and decisions unfold on either side of the border. This is particularly true for the Canada–United States border, which was nearly non-existent 20 years ago. Today, however, the Canada–United States border is very much evident, with much tighter border controls regulating the flow of people, goods, and services—notwithstanding the globalization of the world economy.

1. What, in your view, have been the most significant events of the last 15 to 20 years that have changed the Canada–United States border rules? (Hint: Think about international, national, and even regional events or decisions in politics, economics, and natural resources, etc. that have shaped border controls.)

2. Given the events listed in the previous question, how has a less permeable Canada–United States border affected the everyday lives of Canadian citizens? Provide some specific examples to illustrate.

D. The Politics of Natural Resources

1. There are many reasons for conflict about political boundaries. For example, *territorial disputes* often emerge when a political boundary is superimposed on the landscape and divides ethnically homogeneous populations. *Positional disputes* emerge when bordering states disagree about how a boundary is delimited or about the interpretation of documents that define a boundary. *Functional disputes* emerge when neighbouring states disagree about the policies that exist, or are to be applied, to manage the activities along a border.

   a. Provide a real world example of each of the above types of border disputes. Your examples may be international, national, provincial, or local.

Boundary Dispute	Real World Example
a. Territorial	
b. Positional	
c. Functional	

   b. Some of the most significant boundary disputes concern natural resources. Three prominent examples are (1) Canada's 200-nautical-mile exclusive economic zone off the east coast, where Canada exercises jurisdiction over, among other things, fisheries resources; (2) a disputed region in the Beaufort Sea, where Canada has issued oil and gas exploration permits; and (3) the Northwest Passage in the Arctic, over which Canada claims sovereignty but which other countries view as an international waterway.

   In this exercise you will develop a policy brief on one of the above three resource issues and boundary disputes. The purpose of your policy brief is to inform a newly appointed federal minister of the issue at hand and to identify Canada's policy. Your brief should be no more than one single-spaced page and contain the following:

   i. a clear statement of the boundary dispute
   ii. a description of the region and resource of interest
   iii. a brief background to the boundary dispute, identifying the views of the various parties
   iv. an indication of Canada's current policy or political stance on the boundary issue

2. In 2010 Enbridge Inc., a Canadian crude oil and liquids pipeline company, filed a regulatory application to develop the Northern Gateway Project. The Northern Gateway Project involves the construction of twin pipelines between Bruderheim, AB (northeast of Edmonton) and a marine terminal to be constructed at Kitimat, BC. Each pipeline will be 1,177 km in length. The easterly flow pipeline, from Kitimat to Bruderheim, will carry approximately 193,000 barrels of condensate per day. Condensate is used to thin petroleum products for transport via pipeline. The westerly flow pipeline will transport an average of 525,000 barrels of petroleum per day, from the Athabasca oil sands to Kitimat, where it will be transported to the international market (Asia and the United States). Enbridge estimates that the project will generate $2.6 billion in tax revenue throughout Canada during its operation, and create more than 1,000 long-term jobs. See http://www.northerngateway.ca for additional information about the project.

    Enbridge Inc. argues that the pipeline project is "in the national interest" as it would provide Canada's energy resource sector access to growing Asian and Pacific Rim economies and generate significant economic opportunity. However, some First Nations groups, whose lands would be traversed by the pipeline, as well as environmentalists and other public interest groups have raised concerns about the potential environmental, economic, and social risks associated with the project and argue that the Northern Gateway project is *not* in Canada's national interest.

    a. Political decision-makers reviewing applications for major resource development projects must often make a determination as to whether the project is "in the national interest." A major challenge is that there is little guidance available to decision makers as to what constitutes a "national interest project." In this exercise you will work in groups to develop a list of criteria to assist a decision maker in defining what a national interest project is. Use the above Northern Gateway Project as an example to help you identify potential criteria for making national interest determinations. Identify your criteria in the table below. An example is provided to help get you started.

Principle	Criteria
A natural resource development project is considered to be "in the national interest" when it . . .	a. will help ensure long-term energy security for Canada.
	b.
	c.
	d.

b. Should a natural resource development project be considered "in the national interest" if there is likelihood for significant adverse impacts on local environments, communities or regions? Discuss amongst your group members and explain your reasoning below. Use the Northern Gateway Project as an example to help guide your discussion.

_____
_____
_____
_____
_____
_____
_____
_____
_____
_____
_____
_____
_____
_____
_____
_____

References

Elections Canada. 2008. *Estimation of voter turnout by age group at the 39th federal general election, January 23, 2006.* Ottawa: Elections Canada.

Module 9
The Changing Agricultural Landscape

Introduction

Agriculture is among the most space-consuming activities on the Canadian Prairies. Its geography is shaped by a variety of social, economic, political, and environmental factors. Both the structure and the geography of the agricultural industry, however, have changed considerably in recent decades. In particular, agriculture in Canada has shifted from a very labour-intensive to a very capital-intensive industry. (This is not to say that farming is not still hard work!) Oxen and horses, once used to pull farm equipment, have been replaced by state-of-the-art GPS-guided tractors and combine harvesters.

The agricultural land base itself is also being used much more intensively. Whereas most farms once had chickens, pigs, and occasionally sheep, today's farms are much more specialized in the commodities they produce (Kulshreshtha and Noble, 2007). With increased mechanization, employment in the industry in general is also on the decline, and the traditional "family farm" is slowly becoming a memory.

In this module you will explore the changing agricultural landscape and the geography of food production. First, you will focus on the geography of agricultural production, land use, farm size, and farm practices and on interpreting the driving forces of change and the implications of the changing geography of agriculture. To complete this set of exercises you will be using Statistics Canada's CANSIM interactive socioeconomic database.

You can access CANSIM on the Statistics Canada website (http://www.statcan.gc.ca) or you can access it directly at http://www5.statcan.gc.ca/cansim. In the exercises you will be instructed to compile different data tables by using the CANSIM system. In some cases you will have to download data tables and then generate certain graphs and charts in Excel.

For the last exercise you will visit your local supermarket to collect data on your "food kilometres" and then critically examine the geography of food production and purchasing.

Key Concepts

agricultural production
agriculture intensification
farm cash receipts
farm units
summerfallow
food kilometres

Learning Objectives and Skill Development

1. To understand the geography of agricultural production in Canada.
2. To evaluate shifting geographies of the agricultural industry.
3. To examine the implications of the intensification of agriculture.
4. To evaluate critically the concept of food kilometres.
5. To understand how to work with Statistics Canada's CANSIM database.

Tools Required

Internet access

Note to Instructor

It may be useful to provide students with an overview of how to access and download data from CANSIM. An official CANSIM User Guide may be found here: http://cansim2.statcan.gc.ca/Documentation/GuideOutE.pdf

Student Name: _____

Student Number: _____

Course/Section: _____

A. The Geography of Agricultural Production

In this first exercise you will examine agricultural crop production and production values. To do this exercise you will generate three graphs of agricultural production with the CANSIM tool:

- A line graph of wheat production over time for the prairie provinces
- A pie chart of farm cash receipts for grain crops
- A line graph of farm cash receipts for grain crops over time

Access CANSIM (http://www5.statcan.gc.ca/cansim/) and select "Agriculture" in the "Browse CANSIM by subject" menu.

- Select "Crops and horticulture" and scroll down to select table #001-0017 ("Estimated areas, yield, production, average farm price and total farm value of principal field crops, in imperial units, annual, 1908 to 2013").
- Go to the "Add/Remove data" tab and make the following selections:
    Geography = "Manitoba", "Saskatchewan", and "Alberta". (*Note: be sure to "uncheck" all other boxes*)
    Harvest disposition = "Production (bushels)" (*"Uncheck" all other boxes*)
    Type of crop = "Wheat, all" (*"Uncheck" all other boxes*).
    Time frame = 1908 to 2013
- Screen output format = "HTML table, time as columns"
- Select "Apply"

1. Which province currently produces the largest amount of wheat (bushels)?

   _____

2. Compare 1908 to the most recent year of data. Which province experienced the largest % increase in the production of wheat (bushels)?

   _____

3. Examine the production data for Saskatchewan. In what year(s) was the most significant decline in wheat production? (*Hint: Choose the "Manipulate" tab in the menu above your table and select "Percentage changes, year-to-year" and then "Apply." Alternatively, you can download the data as displayed in the data table tab to Excel and then create a graph*). What key factors might have triggered the decline(s)?

_____
_____
_____
_____
_____
_____

4. What are some of the reasons for fluctuations in wheat production?

_____
_____
_____
_____
_____
_____

Go back to the CANSIM subject menu and this time under "Agriculture" choose "Farm Financial Statistics," and select Table 002-0001 ("Farm cash receipts, annual (dollars), 1971 to 2011").

- Go to the "Add/Remove data" tab and make the following selections:
    Geography = "Canada"
    Type of cash receipts = "Wheat, excluding durum", "Durum wheat", "Oats", "Barley", "Rye", "Flaxseed", "Canola", "Soybeans", and "Corn".
    (*"Uncheck" all other boxes*)
    Time frame = 1971 to 2011
- Screen output format = "HTML table, time as columns"
- Select "Apply"

5. In 2011 (or for the most recent year of available data), what were the largest and smallest grain crop contributors to total farm cash receipts in Canada?

___

6. As measured by farm cash receipts, what crops have shown the greatest increase over the last decade of available data? (*Hint: You can choose the "Manipulate" tab in the menu above your table and examine the "Percentage changes, year-to-year," or you can download the data as displayed in the data table and graph it using Excel.*) Offer an explanation for this.

___

7. As measured by farm cash receipts, what crops have shown the smallest increase (or even decrease) over the last decade of available data? Offer an explanation for this.

___

B. The Changing Practice of Summerfallow

Go back to the CANSIM subject menu. Under "Agriculture" choose "Crops and horticulture" and scroll down to table #153-0039 ("Selected agricultural activities, provinces, every 5 years, (square kilometres), 1971 to 2006").

- Go to the "Add/Remove data" tab and make the following selections:
    Geography = "Saskatchewan" (*Make sure all other boxes are "unchecked"*)
    'Agricultural activities = "Agricultural land area", "Cropland area", and "Improved pasture area" (*"Uncheck" all other boxes*)
    Time frame = 1971 to 2006
- Screen output format = "HTML table, time as columns"
- Select "Apply"
- Record your data in the table below for agricultural land area, cropland area, and improved pasture.

104 | Module 9

Once you have recorded your data, go back to the CANSIM subject menu and select "Agriculture," then "Land use and environmental practices" and scroll down to select table #004-0002 ("Census of Agriculture, total area of farms and use of farm land, Canada and provinces, every 5 years (Number), 1921 to 2011")

- Go to the "Add/Remove data" tab and make the following selections:
  Geography = "Saskatchewan" (*"Uncheck" all other boxes*)
  Total area of farms and use of farm lands = "Summerfallow land" (*"Uncheck" all other boxes*)
  Unit of measure = "Hectares" (*"Uncheck" all other boxes*)
  Time frame = 1971 to 2011
- Screen output format = "HTML table, time as columns"
- Select "Apply"
- Record the data in the table on pages 105–6 for summerfallow area for 1971 to 2011.
- **Note:** Check your units first! 1 hectare = 0.01 square kilometres.

1. Complete the table of agricultural activities in Saskatchewan on the next two pages.

2. What is summerfallow and why is it practised?

   _____
   _____
   _____
   _____
   _____
   _____

3. Describe the overall trend in summerfallow practice as a percentage of total agricultural area over the period 1971 to 2011.

   _____
   _____
   _____
   _____
   _____
   _____

4. What is the percentage change of summerfallow area as a portion of total agricultural area between 1971 and 2006 (the most recent data reported for total agricultural land area)?

   _____
   _____

The Changing Agricultural Landscape

Agricultural activities	Agricultural land area (km²)	Cropland area (km²)	Improved pasture area (km²)	Summer-fallow area (km²)
1971				
1976				
1981				
1986				
1991				
1996				

				2001
				2006
				2011

5. Explain why this trend in summerfallow might be occurring and what the implications might be?

_____

_____

_____

_____

_____

_____

_____

_____

C. Intensification of the Agricultural Industry

Access CANSIM (http://www5.statcan.gc.ca/cansim/) and select "Agriculture" in the "CANSIM by subject" menu. Select "Land use and environmental practices" and scroll down to table #153-0058 ("Selected agricultural activities, Canada, ecozones and ecoregions with agriculture, every 5 years (square kilometres), 1971 to 2006").

- Go to the "Add/Remove data" tab and make the following selections:
    Geography = "Prairies ecozone" (*"Uncheck" all other selections*)

Agricultural activities menu = "Total area", "Number of farm units", "Agricultural land area", "Average farm unit size", "Chemical product expenses", "Chemical product expenses per total area", "Fertilizer expenses", "Fertilized land area", and "Fertilizer expenses per total area" (*"Uncheck"* all *other selections*)
Time frame = 1971 to 2006
- Screen output format = "HTML table, time as columns"
- Select "Apply"

1. The agricultural economy has modified the agricultural landscape in many ways, perhaps the most notable being the change in the total number of farms and the associated change in farm size. Create a bar graph that depicts the "number of farms" and "average farm size" for the period 1971 to 2006. (*Note: You can also select the "Download" tab to download your data an generate your graph in Excel.*)

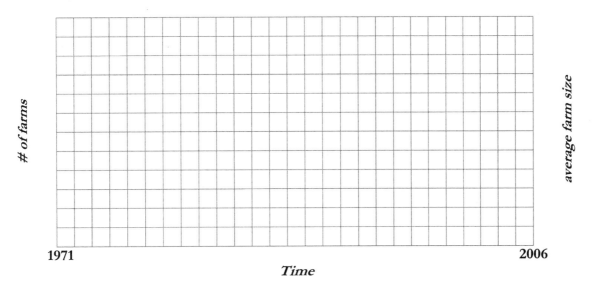

2. Describe the general pattern depicted in the graph above.

___

___

___

___

a. What does this pattern suggest about the intensity of agricultural operations in the Prairie Ecozone?

___

___

___

___

b. What are the implications (social and economic) at the local scale?

3. The graph pattern can be explained in terms of both global and local forces.

   a. What are some of the global forces that might be contributing to the pattern depicted in the graph on the previous page?

   b. What are some of the more local (perhaps farm-level) forces?

4. Using the data you obtained from CANSIM, describe the general pattern of chemical product use and fertilizer use for the time period.

5. Does this pattern seem consistent with the change, discussed previously, in number of farming units and the associated change in farm size? Explain your reasoning. (*Hint: you may want to think about intensive versus extensive agricultural production.*)

_____
_____
_____
_____
_____
_____

6. Aside from the benefits of greater agricultural productivity, what have been some of the other positive as well as negative effects of the increased use of chemical applications and fertilizers?

_____
_____
_____
_____
_____
_____
_____

D. What are your food kilometres?

For most of human history, people ate food that came primarily from local sources. Since advances in transportation and food-processing technology, and with economies of scale in the food industry, the food we eat often travels thousands of kilometres before ending up on our dinner plates! (Halwell, 2002). A recent study by Bentley and Barker (2005), for example, compared the same food products purchased at a supermarket and a farmer's market and found that the imported food at the supermarket travelled an average of 5,364 km compared to only 101 km for local food items. This distance is often referred to as **food kilometres**, and is simply the distance food travels from where it is grown to where it is bought by the consumer. In this exercise you will calculate your food kilometres and examine the merits and demerits of purchasing locally.

1. Visit the supermarket (or shopping centre) where you usually buy your food. Browse the aisles and take an inventory of 20 of the food items you buy most often. Make sure that you include both perishable and non-perishable food.

Note: You are not required to purchase the food items; however, it might save you the extra shopping trip to do so!

Record the information from the product label in the table on the following page. Please note that you can complete column 6 (food kilometres) at home or in the classroom; see the instructions below the table.

Product	Origin*	Brand name (if any)	Unit price (i.e. $ per __)	Perishable? (Y/N)	Food kilometres**
1.					
2.					
3.					
4.					
5.					
6.					
7.					
8.					
9.					
10.					
11.					

12.					
13.					
14.					
15.					
16.					
17.					
18.					
19.					
20.					

*Origin: Indicate the country, state or province, and/or city or town origin of the food product as identified on the product's label. Do not assume that a sign in the store saying "product of . . ." is true. Some labels will give the source of the product, whereas others will give only the place of packaging. For some locally grown foods (e.g., fruits, vegetables, or meats), you may have to ask someone about the source of the product.

**Determine the distance the food product has travelled from its origin to the store where you are shopping. You can use the "geographic centre" of the food's origin as the location from which to calculate distance. For example, if you bought the product at a supermarket in downtown Vancouver and the label says the product was packaged in Ontario (but with no specific address), simply calculate the straight-line distance from the geographic centre of Ontario to your local supermarket as the food kilometres.

2. Complete the chart indicating the percentage of non-perishable food products in the table (pp. 111-12) that are from each origin.

Product origin	Percentage
a. From within your own community or city or the surrounding region	
b. From within your own province	
c. From within Canada	
d. From outside Canada	

3. Complete the chart indicating the percentage of perishable food products in the table (pp. 111-12) that are from each origin.

Product origin	Percentage
a. From within your own community or city or the surrounding region	
b. From within your own province	
c. From within Canada	
d. From outside Canada	

4. For those food products that originate from outside your province or from outside Canada, do they all have local or domestic competitors? Which of the products, as far as you know, do not have local or domestic competitors? Speculate as to why not.

___

___

___

___

___

___

5. What is your "total food kilometres"?

___

6. There has been much discussion in recent years about purchasing food locally. You may have heard of the "100-mile diet" or "eat local, be local" campaigns. What are the advantages and limitations of a move toward buying only food of local (local community or provincial) origin? Consider not only your typical shopping list, but also the geography of food production and environmental impacts.

References

Bentley, S., & Barker, R. 2005. *Fighting global warming at the farmers' market: The role of local food systems in reducing greenhouse gas emissions.* Toronto: Foodshare.

Halwell, B. 2002. *Home grown: The case for local food in a global market.* Danvers, MD: Worldwatch Institute.

Kulshreshtha, S.N., and Noble, B.F. 2007. Agriculture. In B. Thraves, M.L. Lewry, J.E. Dale, and H. Schlichtmann (Eds.), *Saskatchewan: Geographic perspectives* (pp. 337–52). Regina: Canadian Plains Research Centre.

Module 10
Urban Settlement Patterns and Impacts

Introduction

Urban planning is defined by the Canadian Institute of Planners (2010) as the "scientific, aesthetic, and orderly disposition of land, resources, facilities and services with a view to securing the physical, economic and social efficiency, health and well-being of urban and rural communities." Many settlements, both rural and urban, began as planned communities in the sense that their location, focal points, and early street layout were not only determined by influential factors such as proximity to resources and topography, but also by master planning documents. Master plans for Canadian settlements were often altered as each successive generation of residents made different decisions relevant to the politics and values of the day. Today, early settlement patterns are still imprinted at the heart of many cities and towns (e.g. the "old Quebec" or "old Montréal" districts) despite the complex and ongoing generative and degenerative processes of urban renewal and decay.

Original settlement patterns can usually be discerned by examining the layout of the first streets, typically found at the historic town centre or downtown core. In many North American cities, the original town centre and its immediately adjacent neighbourhoods (then the first "suburbs" but now often "inner-city neighbourhoods") were designed on a simple grid pattern of rectangular blocks, arranged row after row in a linear fashion. The "gridiron" street plan dates from antiquity and was used as a measurement system by the Romans. Some of the earliest planned communities displayed a grid street pattern. Many Canadian communities were also influenced by urban settlement patterns in European countries such as France—one of the original colonizing countries. You will examine two such communities later in this exercise. Modern suburbs in North American cities rarely display the traditional design characteristics of early urban settlements, however. Instead, they often adopt a curvilinear design aesthetic primarily meant to slow automobile traffic through residential neigbourhoods and funnel it onto collector roads and freeways. Exceptions can be found in some modern suburbs and communities that reflect "new-urbanism" principles and reintegrate some of the best features of earlier urban forms, including human needs for active transport (e.g., walking or cycling).

The impacts of urban settlement patterns on the environment, society, and the economy are significant and varied, and many human geographers are engaged in understanding and mitigating any negative interactions. For example, if geographers can shed light on the environmental impacts of suburban growth on native prairie swales within city limits, it may lead to changes in municipal sector growth plans. Some geographers are measuring the impacts of suburban design on child obesity rates to try to influence development standards, while others investigate which street patterns lead to improved perceptions of urban quality and vibrant public spaces.

In this exercise, you will learn to identify basic settlement patterns using international examples (Manhattan, USA; Tokyo, Japan; Paris, France; and Brasilia, Brazil) and then apply this knowledge to the analysis of settlement patterns in small Canadian communities. You will also explore modern and traditional suburban design characteristics by performing a comparative field-based analysis of past and present development standards. Finally, you will examine some of the impacts that are associated with different urban settlement patterns.

Key Concepts

settlement patterns
finger pattern
radial pattern
habitat fragmentation

health and environmental impacts
linear pattern

grid pattern
suburban design

Learning Objectives and Skill Development

1. To become familiar with common street patterns in Canadian communities.
2. To understand how urban settlement patterns can impact human health and the environment.
3. To strengthen field observation and data collection skills.
4. To strengthen critical assessment and comparative analysis skills.

Tools Required

Internet access

Student Name: _____

Student Number: _____

Course/Section: _____

A. Common Settlement Patterns

1. Examine the street patterns below and on the following page. Then locate the downtown core of each international city by using Google Maps. Zoom in and out to reveal the patterns as illustrated. Print and submit a screenshot of a section of the downtown core that displays the street pattern in question.

(i) Grid Pattern
e.g., Manhattan, New York

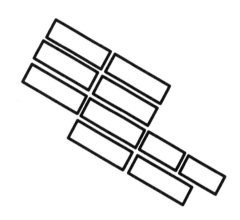

(ii) Radial Pattern
e.g., Paris, France

(iii) Star-Shaped or Finger Pattern
e.g., Tokyo, Japan

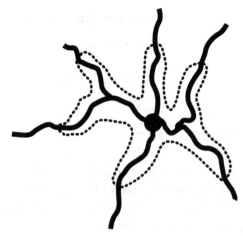

(iv) Linear or Ribbon Pattern
e.g., Brasilia, Brazil

Note: Brasilia was designed to resemble the shape of an airplane. The "wings" of the plane display a linear pattern of development, as does the "boot-shaped" peninsula of land northeast of the "wings."

2. Choose a neighbourhood immediately adjacent to the downtown core in each of the above cities. (a) Is the core model repeated in the neighbourhood? (b) Are any of the other three distinct patterns mentioned above visible in the neighbourhood? (c) What could explain the similarities or differences when comparing the "core" pattern to the patterns in the adjacent neighbourhood?

___
___
___
___
___
___
___
___

3. In Paris, France, what explains the prevalence of the radial design in the core of the city? (Zoom in as close as possible using Google Maps to examine what tends to be located at the centre of the radials.)

_____

_____

_____

_____

_____

4. Examine the downtown district of Tokyo, Japan, using a topographic view in Google Maps.

   a. How is the star-shaped (or finger) design influenced by the topography of the island?

   _____

   _____

   _____

   _____

   _____

   b. Based on what you found for Tokyo, in what region(s) of Canada might the star-shaped (or finger) pattern be prevalent in cities?

   _____

   _____

   _____

   _____

   _____

B. Settlement Patterns in Canadian Communities

Now that you are familiar with the urban models and their international examples, in this exercise you will try to match the settlement patterns of several small Canadian settlements with one of the urban models provided earlier to determine which model best represents the pattern at the centre of each small community. Again using Google Maps as your tool, examine each of the following small Canadian communities. You should be able to see the entire community within the viewing frame. This will allow you to discern the settlement pattern more easily than by viewing the community from either too close up or too far away.

The communities are:

- Humboldt, Saskatchewan
- Oliver, British Columbia
- Forestville, Quebec
- Charlesbourg (and nearby Bourg-Royal), Quebec

1. Which common settlement pattern (introduced in Question A.1.) is reflected in the street design at the core of each of these smaller communities?

Canadian Community	Closest Representative Urban Model
Humboldt, Saskatchewan	
Oliver, British Columbia	
Forestville, Quebec	
Charlesbourg (and nearby Bourg-Royal), Quebec	

2. Now obtain a map of the community you are living in. Are any of the urban models provided earlier reflected in the core settlement pattern of your community? Describe any similarities or differences that you see.

_____
_____
_____
_____
_____
_____
_____
_____
_____
_____

3. (a) Provide a brief assessment of the potential merits and demerits of each of the common settlement patterns in the table below based on i) ease of transportation (public bus service), ii) relative cost of municipal servicing (e.g., cost to install and maintain utility infrastructure), and iii) liveability (i.e., from the perspective of a resident pedestrian or cyclist).

Settlement Pattern	Characteristic		
	(i) Ease of transportation	(ii) Relative cost of municipal servicing	(iii) Liveability
Grid			
Radial			
Star-shaped (Finger)			
Linear (Ribbon)			

(b) Which pattern would you recommend your community adopt and why?

_____

_____

_____

C. Comparing Past and Present Suburban Development Standards: Impacts to Human Health

Visit one of the original residential areas in your community, preferably just outside the central business district. Select one representative block within this neighbourhood as your study area. Now take a number of measurements based on the table (page 123) and record your observations as you move about the study area. Repeat this exercise in one of the newest residential areas in your community, again choosing a representative block in the neighbourhood as your study area. You will need some extra sheets of paper on which to perform rough calculations and jot down field notes. Refer to Module 1 for a refresher on how to estimate distances in the field. You might also find it useful to bring along a digital camera to take field photos of the streetscape. This will help you recall the details of your streetscape as you write up your results. If you are unsure if a space is public or private, consult with your instructor. Do not traverse private property unless you have consent from the property owner. If a property owner is absent, err on the side of caution and do not trespass.

1. In the table on the next page, record the development characteristics of each of your selected residential areas.

Urban Settlement Patterns and Impacts | 123

	Original Residential Area	Modern Residential Area
Measurements (in metres)		
Width of road		
Estimated average depth of front yard		
Estimated average distance between housing units		
Estimated average width of single-family housing units measured at ground floor level		
Observations		
Does this suburb feature a curvilinear or linear street pattern?		
How many multi-unit dwellings are in this block?		
How many garage doors face the street?		
How many porches or stoops face the street?		

2. Summarize your data by writing a one-page comparative analysis of the two neighbourhoods. Discuss common and unique design characteristics and comment on how suburban design standards in your community appear to have changed over time.

3. For in-class debate: (a) Which type of suburban area provides a higher standard of urban quality for citizens (where urban quality is defined as the degree of human comfort, safety, and pleasure made possible by the physical features of an area)? (b) Prepare to defend your answer by recording four supportive points in the space below.

___

___

___

___

___

___

___

4. According to the Harvard School of Public Health, the built environment (streets, parks, pathways, etc.) plays a significant role in making physical activity a daily habit. Which neighbourhood design do you feel better supports human well-being (avoidance of obesity being just one aspect)? Explain.

___

___

___

___

___

___

___

D. Urban Growth and Impacts to Natural Areas

Urban settlements also impact the natural environment. For example, in Canada's prairie provinces, swales featuring native prairie ecosystems are at risk of being degraded or destroyed as suburban development consumes land within city limits. A swale is a low-lying tract of land that is especially moist or marshy. Swales are often dotted with ponds and are highly attractive as habitat to a diversity of terrestrial, aquatic, and avian wildlife species. The cumulative effects of road development can be particularly devastating to swales if they fragment relatively intact patches of native grasslands. To help municipal land use planners decide where to locate new roads and prevent the negative effects of habitat fragmentation, human geographers can calculate "effective mesh size" and "effective mesh density."

The "effective mesh size" metric expresses the probability that two points chosen randomly in a region are connected—that is, they are not separated by barriers such as roads (e.g., points at the north and south end of an intact patch of swale within city limits). The more barriers (roads) fragmenting the landscape (the swale), the lower the probability that the two points are connected and the lower the "effective mesh size." A lower "effective mesh size" also means there is a lower probability that animals or people are connected across the same landscape due to landscape fragmentation (i.e., the lower the possibility of meeting each other without having to cross a barrier). In an area like a swale, this may mean decreased opportunities for wildlife mating, increased wildlife mortality (road kills), and interruption of seasonal wildlife migration patterns.

To make it possible to compare values from various areas, the probability of being connected is converted into the size of a patch—"the effective mesh size"—by multiplying it by the total area of the region investigated. The "effective mesh size" is expressed in square kilometres. It is calculated as follows:

$$m_{eff} = p \times A_{total} = \left(\left(\frac{A_1}{A_{total}}\right)^2 + \left(\frac{A_2}{A_{total}}\right)^2 + \left(\frac{A_3}{A_{total}}\right)^2 + \cdots + \left(\frac{A_n}{A_{total}}\right)^2\right) \times A_{total}$$

or, in brief:

$$m_{eff} = \frac{1}{A_{total}} \sum_{i=1}^{n} A_i^2$$

where n represents the number of patches, A_1 to A_n the patch sizes from patch 1 to patch n, and A_{total} the total area of the region investigated.

The first part of the formula gives the probability p that two randomly chosen points (for example, in a swale) are in the same patch. The second part (multiplication by the size of the region, A_{total}) converts this probability into a measure of area. This area is the "mesh" size of a regular grid pattern showing an equal degree of fragmentation and can be directly compared with other (possibly more intact) regions (of swale).

Instead of only talking about the size of meshes, it is also possible to talk about the number of meshes in an area. The "effective mesh density" gives the effective number of meshes per 1,000 km²—in other words, the density of meshes. This is very easy to calculate from the "effective mesh size": it is simply a question of how many times the effective mesh size fits into an area of 1,000 km². It is calculated as follows:

$$s_{eff} = \frac{1000 \text{ km}^2}{m_{eff}} \times \text{meshes per } 1000 \text{ km}^2$$

The mesh density value rises if fragmentation increases. So, the two measures contain the same information about the landscape, but the mesh density is more suitable for spotting trends because it illustrates what is going on (e.g., an increasing value demonstrating increases in fragmentation caused by roadways in and around cities).

Imagine that an intact prairie swale within city limits is fragmented by roads into three patches:

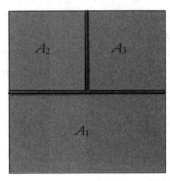

$A_{total} = 2 \text{ km} \times 2 \text{ km} = 4 \text{ km}^2$

1. Calculate the probability that two randomly chosen points in the swale will be in patch 1 and therefore connected. To do so, first determine the probability that one randomly chosen point will land in patch 1. Then determine the probability that a second randomly chosen point will land in patch 1. The probability that both points will be in patch 1 is the product of these two numbers. (Do you recognize this product as part of the formula of effective mesh size indicated previously?)

2. Calculate the probability that two randomly chosen points in the swale will be in patch 2 (the answer for patch 2 will be the same for patch 3 since they are the same size). Follow the same idea as in the previous step (but consider that patch 2 has a different size than patch 1).

Urban Settlement Patterns and Impacts | **127**

3. Calculate the probability that the two points will be in patch 1, patch 2, or patch 3, by finding the sum of the three probabilities. Record your answer here: _____.

4. Multiply this probability by the total area of the region to find the value of the "effective mesh size." Record that value here: _____.

5. Calculate the "effective mesh density," using the formula given earlier.

6. Now re-calculate "effective mesh size" and "effective mesh density" assuming a total area of 25 km² (and thus a more intact swale area, i.e., larger patches).

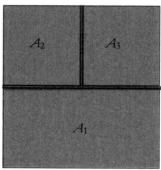

A_{total} = 5 km × 5 km = 25 km²

7. Imagine that a provincial government is planning to approve the construction of a new highway inside city limits through a relatively intact swale that parallels a protected riverbank buffer zone (managed by a river valley protection authority). The swale is already bisected by an existing (partially completed) municipal ring road system. Based on what you have learned about habitat fragmentation from the previous calculations, write a short statement to the provincial government advising it about the impacts of the freeway development. Black bears, white tailed deer, and muskrats are known to frequent both the swale and the riverbank buffer zone.

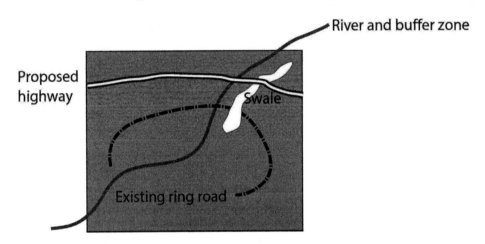

References

Canadian Institute of Planners. 2010. "Planning is…" www.cip-icu.ca/web/la/en/pa/3FC2AFA9F72245C4B8D2E709990D58C3/template.asp. Retrieved on 14 May 2010.

Harvard School of Public Health. 2012. "Environmental barriers to activity." http://www.hsph.harvard.edu/obesity-prevention-source/obesity-causes/physical-activity-environment/

Jaeger, J. 2000. Landscape division, splitting index, and effective mesh size: new measures of landscape fragmentation. *Landscape ecology* 15(2): 115–130.

Jaeger, J., Bertiller, R., and Schwick, C. 2007. "Degree of landscape fragmentation in Switzerland: Quantitative analysis 1885–2002 and implications for traffic planning and regional planning." Condensed version. Federal Statistical Office, Neuchâtel.

Module 11
Urban Economies and Transportation

Introduction

Fundamental to all cities and settlement regions is economic activity—the manufacture, exchange, and consumption of goods and services. One of the economic aspects of urban life explored by North American human geographers is the link between capitalist modes of production and urbanization. Closely intertwined with this subject is the role of transportation and distance to market.

Building on these broad themes, in this series of exercises you will explore factors affecting the push and pull of economic development in a city, investigating both how travel distance is related to the diversity of economic function in a retail centre and how the attractiveness of a particular economic area is related to the overall quality of streetscape characteristics.

Key Concepts

retail gravitation	degree of attractiveness	Huff model
economic profile	urban transect survey	network connectivity

Learning Objectives and Skill Development

1. To calculate retail attractiveness for competing retail destinations.
2. To assess critically the Huff model of retail attractiveness.
3. To apply a spatial method to learn the characteristics of a population or place.
4. To develop awareness of indicators of economic activity and the "degree of attractiveness" of economic centres.
5. To learn to assess the quality of a streetscape environment by using a standard scale rating.

Tools Required

No special equipment is needed.

Student Name: _____

Student Number: _____

Course/Section: _____

A. Urban Retail Markets

How many retail outlets are needed in a particular urban market area? Where should we locate retail outlets to best serve the urban market? How is the urban market best divided amongst competing retail locations? These questions are fundamental to understanding urban economic systems and to ensuring that urban market areas—as well as the retail and service centres that comprise them—are sustainable.

Geographers can express the retail sales expected at a particular location in an urban market area as a function of the attractiveness of the goods and services provided (e.g., diversity, quality, and square footage of selling space as an indicator of the attractiveness of the retail centre), the amount of demand in the market area, and the distance (or cost of distance) associated with consumer travel to the retail location. For any single retail location, however, its market area and sustainability depend not only on the goods and services provided and the accessibility of its location, but also on the characteristics and locations of its competitors. These variables, which affect the sustainability of a retail location and determine, in part, the size of its market area can be simplified and expressed in terms of probabilities. The *Huff model* is one such approach, measuring the probability that a consumer at location i will shop at retail location j. The *Huff model* suggests that the attraction of a shopping centre or store in a metropolitan area depends on the size of the shopping centre or store (e.g., product assortment, total floor space), distance, and sensitivity to time. Simplified, the Huff model consists of five parameters:

- $P_{(ij)}$ is the probability that a consumer travelling from place i will shop at place j.

- S_j is a measure of the attractiveness of the shopping centre or store j. An example of this might be the number of retail outlets in a big box area, or the total floor space of a large shopping centre or supermarket devoted to a particular type of product.

- D_{ij} is the distance from a consumer's place of origin to the shopping centre or store location. But, in urban centres the travel time is often more meaningful than distance, and this parameter is expressed as T_{ij}, or the travel time from the consumer's place of origin i to the shopping centre or store location j.

- λ is a calibration of the model to estimate the effect of travel time on different kinds of shopping trips (e.g., big single-item purchases or multi-purpose shopping trips).

Simply put:

$$P_{(ij)} = \frac{S_j / (T_{ij})^\lambda}{\sum_{j=1}^{n} \frac{S_j}{(T_{ij})^\lambda}}$$

For the purposes of this example, we will assume five shopping centres, each with similar products and services being offered, and four city neighbourhoods. For simplicity, we will assume that $\lambda = 1$.

Shopping centre j	Selling space (centre size) (S_j)*
Delaware Centre	3.8 units
Richmond Centre	2.2 units
Bayview Outlets	1.8 units
Central Centre	4.2 units
Riverside Centre	3.4 units

* 1 unit of selling space = 100,000 square metres

Neighbourhood (i)	Average travel time (minutes) from neighbourhood to shopping centre (T_{ij})				
	Delaware	Richmond	Bayview	Central	Riverside
Yaletown	23	12	15	31	6
Clareview	43	12	18	25	14
Brentwood	40	24	11	18	13
Southlands	10	23	16	22	4

1. Calculate the probability that a consumer from neighbourhood i will shop at each of the shopping centres j.

Neighbourhood (i)	Probability of shopping at each centre				
	Delaware	Richmond	Bayview	Central	Riverside
Yaletown					
Clareview					
Brentwood					
Southlands					

2. Assume a developer has submitted an application for a new shopping centre. How might the developer or a city planner use the information in the previous table to help make a decision about the proposed shopping centre? (i.e., for what purposes can this information be used?)

3. Models are, of course, simplifications of reality.

   a. What other information would be required about retail patterns or urban market areas to help make an informed decision about where to locate a newly proposed shopping centre?

   b. In the example in question 1, the model was applied to large shopping centres. Would this model be equally applicable to small specialty stores, such as those selling only one type of product? How about coffee shops, such as Tim Horton's or Starbucks? Explain your reasoning.

B. Network Connectivity

A determining factor of the propensity for consumers to shop at any given location is the connectivity of the shopping centre within the urban market area. The connectivity of a location depends on the characteristics of the transportation network. A transportation network is simply a system of routes (sometimes referred to as edges) and places (referred to as vertices or nodes). *Network connectivity* refers to the case of movement from one node to another within a given system of routes and can be determined by the Beta index (ß):

$$\beta = \frac{E}{V}$$

where,

E is the number of edges in a network

V is the number of nodes.

When ß is large, the network is "well connected." When ß is small, the network is "poorly connected."

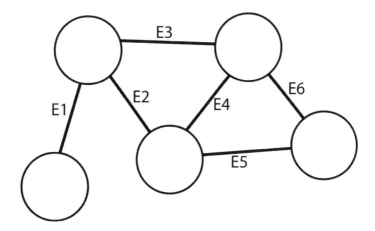

1. Calculate the Beta index for each of the following networks:

   a. 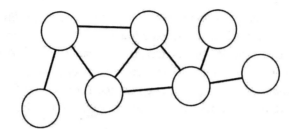 ß = _____

   b. 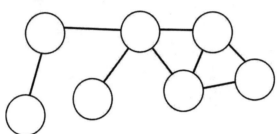 ß = _____

   c. 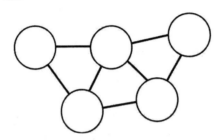 ß = _____

   d. 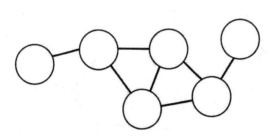 ß = _____

2. Which of the above transportation networks is the "most connected"? _____

3. (a) Other than measuring the accessibility of a retail location, why would geographers or urban planners be interested in knowing the connectivity of a transportation network? (b) Provide three examples in an urban setting for which understanding network connectivity would be important.

   _____
   _____
   _____
   _____

C. Main Street Economic Profile

In this exercise, you will perform a transect survey to create an economic profile of the main street, or commercial business strip, in your city or town. The transect survey is a spatial method for determining key characteristics about a population region based on observing only a small area that "cuts across" the area of interest in either a "line" or "strip" fashion. This method is commonly used in environmental sciences and social sciences, including human geography. The observations recorded along a transect are then used to estimate the characteristics of the entire area. In this kind of survey, all observations that fall outside the designated strip must be ignored.

In archaeological field studies, for example, line-transect surveys can be used to gather data on the location and density of artifacts or other materials. Small research teams might work together to walk selected transects (ignoring topography). In this case, transects in remote areas are often determined (or "made visible") with the help of a GPS unit. GPS is especially helpful if it is important for the team to take samples at regular intervals.

In human geography, urban planners use transect surveys in a slightly different way. If, for example, a research team was interested in gathering data about how urban design may affect health in various neighbourhoods, they might do a transect survey in each neighbourhood of interest. In this case, they would select a strip transect within each neighbourhood and gather data there, rather than survey the whole neighbourhood. This is the type of exercise you are about to undertake.

Identify a prominent commercial district or strip in the downtown area of your town or city (perhaps Main Street or the "high" street). Using the length of your average pace (see Module 1), you will survey three 100-m transects (roughly equivalent to one city bock), **each on a different street**.

For each transect, choose a starting point and briskly pace off 100 m. Then return to the starting point of each transect and walk its length more slowly, observing the economic activity and land use present. Record your observations in the table on the next page under the categories provided. Later, you will use these data to answer the questions C.1.–C.4.

It is suggested that you work in groups of three or four and assign each person responsibility for recording certain observations in the table. Make sure to copy all information collected by the group to your own table before submitting the assignment.

1. What is the ratio of the total number of businesses (i.e., the number of businesses counted) to the total number of business functions (i.e., the number of business functions or categories present) for each of the following:

	Transect 1	Transect 2	Transect 3
a. For each transect			
b. For all three transects			

2. What is the ratio of the total number of businesses (i.e., the number of businesses counted) to the total number of buildings for each of the following:

	Transect 1	Transect 2	Transect 3
a. For each transect			
b. For all three transects			

3. What is the ratio of national and international to local businesses for each of the following:

	Transect 1	Transect 2	Transect 3
a. For each transect			
b. For all three transects			

4. Write a brief description, based on your data, of the economic conditions and level of development of your streetscape; note and explain any differences between transects.

   _____
   _____
   _____
   _____
   _____
   _____
   _____
   _____

Businesses Observed	Transect 1 (100 m)	Transect 2 (100 m)	Transect 3 (100 m)	Transect totals	
Manufacturing					Tally the number of businesses along each transect according to the "business function" categories below. Note that one building may have multiple businesses. Be sure to count all businesses.
Transportation & warehousing					
Trade					
Finance & insurance					
Real estate & leasing					
Professional, scientific, technical services					
Business services					
Educational services					
Health services					
Information, culture & recreation					
Hospitality & food services					
Public administration					
Other services					
total # of buildings					
total # of entrances that lead to an active (not vacant) establishment					
total # of national vs. international businesses						
total # of local businesses					
total # of vacant, shut down, or unoccupied buildings or building units					
total # of parking lots/undeveloped spaces					

D. Main Street: Degree of Attractiveness

Now you will consider your main street transect from another perspective: rating its overall "degree of attractiveness" from an urban quality perspective (where urban quality is defined as the degree of human comfort, safety, and pleasure made possible by the physical features of an area). The overall degree of attractiveness of a commercial street can strongly influence the economic success of the business owners on a street. An "attractiveness" rating scale is provided in the table on page 140 (Gehl Architects, 2004). While this scale is largely subjective and based on personal perceptions of the streetscape, subjective (qualified) observations can be just as important as objective (quantified) observations when one is trying to determine what changes might increase the economic success of a commercial area. "Degree of attractiveness" rating scales have been used by architects and city planners in cities such as Sydney, Australia; New York City; and Saskatoon, Canada, to inventory attractive commercial streets and determine how less successful streets might be made more attractive.

Familiarize yourself with the "degree of attractiveness" rating scale on page 140. Some of the criteria will be the same as those in the previous exercise, but the criteria in the rating scale also include factors related to the design of ground-floor units. Then take another look at each of your main-street commercial blocks, paying close attention to the quality and condition of ground-floor building frontages.

1. Using the rating scale and description provided in the table on page 140, what rating should each of your transects receive?

    Transect 1: _____

    _____

    Transect 2: _____

    _____

    Transect 3: _____

    _____

2. (a) Do you agree with these ratings? Why or why not?

    _____

    _____

    _____

    _____

    _____

    _____

    _____

(b) How would you adjust the scales to make them more representative of the features of the built environment in the area you are investigating? Be specific.

_____
_____
_____
_____
_____
_____
_____

Rating	Description
A = Attractive	Small units (15–20 units per 100 metres) Many doors or entrances to units Diversity of functions Few or no closed or unoccupied units Quality materials Interesting relief or texture in facades No building setback from sidewalk or street
B = Pleasant	Relatively small units (10–14 per 100 metres) Many doors or entrances to units Some diversity of functions Only a few closed or unoccupied units Some relief in the facades Relatively good detailing Shallow or no building setback from the sidewalk or street
C = Somewhere in between pleasant and dull	Mixture of small and larger units (6–9 units per 100 metres) About the same number of doors as units Some diversity of functions Possibly several closed or passive units Uninteresting facade design Somewhat poor detailing Shallow or moderate building setback from the sidewalk or street
D = Dull	Larger units with few doors (1–5 units per 100 metres) Little diversity of functions Many closed and unoccupied units Predominantly unattractive facades Some zero architecture facades (e.g., parking lots) Moderate building setback from the sidewalk or street
E = Unattractive	Units are predominantly large with few or no doors No visible variation of function Monotonous facades Predominantly unattractive or industrial building materials (e.g., concrete) No details, nothing interesting to look at Many closed units and/or zero architecture locales. Buildings may be separated from the street by parking lots
F = Repellent	Like E, but even more uninviting Devoid of visual interest Unkempt or abandoned

3. What other key factors or indicators, which perhaps are not reflected in the scale, affect *your* overall assessment of attractiveness?

4. Using the rating scale on the previous page, how would you rate the attractiveness of a typical North American big-box commercial development strip, and why?

5. Does the scale work for suburban retail areas such as big-box malls? How might it be adapted to suit this context?

References

Gehl Architects. 2004. *City to waterfront: Wellington public spaces and public life study*. Wellington City Council: Wellington, New Zealand.

Module 12
Understanding the Industrial Landscape

Introduction

Why are industries located where they are? Why do certain regions specialize in certain types of economic activities? How is the industrial landscape changing in an era of globalization? What are the effects on local economies and livelihoods? These are some of the fundamental questions that human geographers are tackling concerning the evolving industrial landscape in Canada and internationally.

To understand patterns on the industrial landscape it is important to understand the changing divisions of labour. Under the traditional Fordist industrial system, for example, labour and production were divided internally and technically—often within the firm itself and also at the local to regional scale (e.g., steel towns, coal towns, etc.).

Although such steel towns and coal towns still exist, as a result of globalizing tendencies there has emerged a new spatial organization of the labour force, termed the **new international division of labour**. This term refers to the emerging worldwide division of labour associated with the globalization of production. For example, increasingly raw materials (e.g., steel and other metals) are extracted and partially processed in one country, shipped around the globe to where they are transformed into parts of product (e.g., automobile parts), and transported again to an assembly plant where the parts are assembled into a final product. The final product is sometimes shipped back to the origin of the raw material, and to other places, where it is sold to the market.

In this module you will examine the implications of the changing industrial landscape on both the industrialized and industrializing world. In the first part of this module you will examine Canada's industrial landscape, particularly its labour force, and shifts in the labour force over time and by industrial sector. Attention then turns to industrial location, particularly of manufacturing industries, and to understanding the implications of export-processing zones. Finally, the concept of economic multipliers is introduced, and you will learn how to predict the impact of new industry start-up and industry relocation on a regional labour force.

Key Concepts

labour force classification
services-producing sector
disappearing middle

goods-producing sector
export-processing zone
economic multipliers

location theory
industry classification

Learning Objectives and Skill Development

1. To examine changing patterns in the Canadian labour force.
2. To extract and analyze labour-force data from Statistics Canada's CANSIM database.
3. To critically evaluate traditional location theory in geography.
4. To research and evaluate the nature and impact of export processing zones.
5. To apply economic multipliers to assess the impact of industrial location decisions on the local labour force.

Tools Required

Internet access

Understanding the Industrial Landscape | 145

Student Name: _____

Student Number: _____

Course/Section: _____

A. The Changing Canadian Labour Force

In this first exercise you will be using the CANSIM database on the Statistics Canada website (http://www.statcan.gc.ca), introduced in Module 9. You can access CANSIM directly at http://www5.statcan.gc.ca/cansim/. First, you will examine the trend over time in national employment in the goods-producing sector and the services-producing sector. Then you will examine in more detail employment in various goods- and services-producing sectors in Canada and in your home province.

Go to CANSIM and select "Labour" in the "Browse CANSIM by subject" menu. Select "Industries" from the list, and then scroll down to select Table 282-0008, "Labour force survey estimates (LFS), by North American Industry Classification System (NAICS), sex and age group, annual (persons), 1976 to 2012."

- Go to the "Add/Remove data" tab and make the following selections:
  Geography = Canada
  Labour force characteristics = "Employment"
  North American Industry Classification Systems = "Goods producing sector" AND "Service producing sector" (*"Uncheck" all other selections*)
  Sex = "Both sexes"
  Age group = "15 years and older"
  Time frame = 1976 to 2012
- Screen output format = "HTML table, time as columns"
- Select "Apply"

1. Record in the table below the total labour-force employment in the goods-producing industries and services-producing industries for the dates specified.

	1976	1982	1988	1994	2000	2006	2012
Goods-producing sector							
Services-producing sector							

2. Describe the trend in labour-force employment in the goods-producing industries and services-producing industries between 1976 and 2012. Speculate as to what might be causing this trend.

_____
_____
_____
_____
_____
_____

Go back to the CANSIM subject menu. Select "Labour" and then "Employment and unemployment." Scroll down to select Table 282-0008.

- Go to the "Add/Remove data" tab and make the following selections:
    Geography = Canada
    Labour force characteristics = "Employment"
    North American Industry Classification System =
        "Total, all industries"
        "Agriculture"
        "Forestry, fishing, mining, quarrying oil and gas"
        "Manufacturing"
        "Trade"
        "Transportation and warehousing"
        "Finance, insurance, real estate and leasing"
        "Professional, scientific and technical services"
        "Educational services"
        "Health care and social assistance"
        "Information, culture and recreation"
        "Accommodation and food services"
    Sex = "Both sexes"
    Age group = "15 years and older"
    Time frame = "1976" to "1976," scroll down, and choose "Retrieve as a table."
- Screen output format = "HTML table, time as columns"
- Select "Apply"

3. Calculate the labour force percentage for each industry class (out of the "Total, all industries") and record your values for "Canada 1976" in the table on the following page.

    - Repeat the above procedure, but for the "Reference period" choose "2010" to "2010." (*Hint: You can use the back button on your browser and under "Retrieve data for the following time period" indicate "2010" to "2010."*)

4. Calculate the labour force percentage for each industry class (out of the "Total, all industries") and record your values for "Canada 2012" in the table below. (*Hint: To obtain the data for 2012 select the "Add/Remove" data tab and scroll down to change the "Time frame."*)

5. Calculate the labour force percentage for each industry class (out of the "Total, all industries") and record your values for "YOUR OWN PROVINCE 2012" in the table below. (*Hint: To obtain the data for your own province select the "Add/Remove" data tab and scroll down to change the "Geography."*)

Industry classification	Canada, 1976 (employment, %)	Canada, 2012 (employment, %)	Your province, 2012 (employment, %)
Agriculture			
Forestry, fishing, mining, oil, and gas			
Manufacturing			
Trade			
Transportation and warehousing			
Telecommunications			
Finance, insurance, real estate, and leasing			
Professional, scientific and technical services			
Educational services			
Health care and social assistance			
Information, culture, and recreation			
Accommodation and food services			

6. What were the *top three* and *bottom three* industry sectors in percentage of labour-force employment in Canada in 1976 and 2012?

	1976	2012
Top three industry sectors		
Bottom three industry sectors		

7. What sector(s) had the largest percentage increase in employment between 1976 and 2012 in Canada? What sector(s) had the largest percentage decrease in employment between 1976 and 2012? Do you notice any patterns in terms of goods-producing sectors versus services-producing sectors?

_____
_____
_____
_____
_____
_____

8. Compare your province to all of Canada in 2012. Are there industry sectors in your province for which the percentage of labour-force employment is significantly above or significantly below the national statistic? If so, can you explain why these differences might exist (i.e., why does your province specialize, or not specialize, in certain activities)?

_____
_____
_____
_____
_____
_____

9. It has been said, both nationally and internationally, that we are witnessing a *disappearing middle* in the labour force. In other words, the number of high-paid professional jobs is increasing, as are the number of relatively low-paid service jobs, and there are fewer "middle-class" or blue-collar jobs. Do the statistics above suggest that this trend appears to be evident in the Canadian economy? Explain.

10. What do you think are some of the forces that are contributing to the phenomenon of a disappearing middle? What are the potential implications?

B. Industrial Location

Alfred Weber (1868–1958), the German economist, is well known for his general theory of the location of industry. Weber's theory of least-cost industrial location does not describe where industries are actually located on the landscape; rather, it prescribes where industries *ought to be* located in light of a number of factors related to transportation, raw material locations, labour, and markets. Assuming that labour is not mobile and markets are geographically fixed, Weber's theory of industrial location suggests that bulk- and weight-reducing industries should be located near the raw-material site.

Today, however, most manufacturing plants begin the manufacturing process with semi-finished components rather than raw materials, and transportation costs for shipping raw materials remain relatively low in comparison to shipping rates for finished products. Consider, for example, how much shipping you pay for a courier to deliver a single product that you purchased online. Many finished products are also much smaller and lighter today than they were 20 to 30 years ago. Just ask your parents or grandparents about their floor-model television, vinyl, or 8-track players. (No, the latter two are nothing like iPods!)

1. Take an inventory of products in your classroom, your lab, or your home, and note where those products are manufactured. If you are doing this exercise in the classroom, work with your classmates to compile this information. Document your inventory in the table on the following page.

2. In what country is the majority of the products you identified manufactured?

_____

_____

Category	Specific item	Manufacture location
Textiles (e.g., footwear, clothing, other fabric items)		
Electronics (e.g., iPods, calculators, cellular phones, projectors)		
Stationery (e.g., notepads, pens, binders, whiteboard pens)		
Furniture (e.g., chairs, tables, desks)		

3. Is there a geographic clustering of where certain types of products are manufactured? If so, can you explain why? (*Hint: Consider the factors that affect industrial location.*)

___

C. Export-Processing Zones

In an era of globalizing industry, **export-processing zones** (EPZs) have emerged as enclaves of industrial development. Currently there are EPZs in more than 100 countries! EPZs are areas of a country where typical trade and business investment barriers are often relaxed, or where certain investment and business-development incentives are provided so as to attract foreign investment and business. EPZs are usually labour-intensive, export-based manufacturing zones, most of which are located in developing countries (e.g., Malaysia, Indonesia, Pakistan, Mexico, Costa Rica, Honduras, Namibia).

In this exercise you will explore the nature and characteristics of EPZs, and assess critically both their positive and adverse effects for the host country. First, use the Internet to find an EPZ in a developing nation, create a profile of the EPZ based on the characteristics listed in the table on the following page, and then answer the questions that follow.

1. List the characteristics of your EPZ using the table on the following page.

Export processing zone	
Host country	
Local geography (coastal or inland?)	
Website address	
Types of industries or exports	
Benefits offered to multinational and transnational corporations	
Does the EPZ offer any special terms or conditions to foreign investors?	

2. Why do you suppose most EPZs are located in developing countries?

3. One advantage that many EPZs offer is a coastal location. Why do you suppose this is important for attracting international industry and investment?

4. What are the potential benefits of export-processing zones to a host country?

5. What are the potential adverse effects of export-processing zones to a host country?

D. Employment Multipliers

When new industries locate to a region they generate not only new income but also new employment. When analyzing the potential impact of a new industry, geographers often make use of multipliers. A **multiplier** summarizes the total impact that can be expected from a change in a given economic activity. The **employment multiplier** is a particular type of multiplier that measures the total change in employment resulting from an initial change in employment of an exporting industry—the economic base or basic industry of a region.

When basic industries grow, they create new jobs in other, non-basic industries. For example, if a region specializes in automobile assembly and a new assembly plant is located in the region, there will be a contribution to the regional economy not only of basic jobs (i.e., jobs in the new assembly plant), but also jobs in other, non-basic industries. The multiplier for a basic industry provides an estimate of the total new jobs that will be created.

A regional multiplier (k) can be defined as total regional employment (E_t) divided by regional employment in basic jobs (E_b). Expressed another way, total regional employment equals regional employment in basic jobs multiplied by a regional multiplier, or $E_t = kE_b$.

1. Region A specializes in the mining and metals industry. The total labour force in Region A is 5,800 people, of which 1,800 work at basic jobs and 4,000 at non-basic jobs. What is the multiplier for Region A?

2. New mineral deposits are discovered and new mining investors are attracted to the region. A new mining operation is established, employing 520 new people in basic jobs. How many total new jobs can be expected in Region A?

3. A new minerals-processing plant is being proposed, and two mining regions are competing to attract the new growth in their basic industry. Mining Region X has a total workforce of 2,400, of which 800 work at basic jobs. Mining Region Y has a total workforce of 3,550 of which 980 work at basic jobs. The new processing plant will create 225 new basic jobs. If the goal of the processing plant's owner is to maximize its impact on the regional economy, where should it choose to locate the new plant? (*Hint: First calculate each region's multiplier.*)

4. Assume that the processing plant is relocating. Currently, the processing plant employs 310 workers in Region Z, which is also a mining and metals region. If Region Z has a multiplier of 2.3, will the new job creation in Region X or Y offset the total employment loss in the industry in Region Z?

Notes